Experimental
Biology
A Laboratory Manual

Experimental Biology

A Laboratory Manual

Abhijit Dutta

Alpha Science International Ltd.

Oxford, U.K.

Experimental Biology
326 pgs. | 65 figs.

Abhijit Dutta
Department of Zoology
Ranchi University
Ranchi, India

ALPHA SCIENCE INTERNATIONAL LTD.
7200 The Quorum, Oxford Business Park North
Garsington Road, Oxford OX4 2JZ, U.K.

www.alphasci.com

ISBN 978-1-84265-432-3

Printed in India

DEDICATION

The book is dedicated to our beloved mother, Smt. Sandhya Dutta, who left us for her heavenly abode on 5th August 2007, whose constant encouragement and impetuos enabled us to complete the book

PREFACE

Spectacular advances have taken place in science and technology in the last century. Keeping this in mind UGC has introduced new practical curricula for undergraduate and postgraduate students of various universities. It is thus mandatory that infrastructure facilities in life science should be provided in different colleges and universities so that our students should be able to compete nationally and internationally. For reasons not known there lies a wide range of disparity between the theoretical and practical studies, more so with the introduction of new UGC syllabus. As a result both teachers and students of undergraduate and postgraduate level face tremendous difficulties in acquiring study material of practical utility, particularly topics related to experimental, analytical and separation techniques.

We though it reasonable, therefore, to write *"Experimental Biology"*, which presents both the principles of analytical biology that makes experimental, analytical and separation techniques possible in the laboratory, with background information to understand what the students are going to do and why, including technical information the students gather in the laboratory. The book is subdivided into different parts, including Chapters like: Immunology, Microbiology, Animal Cell Culture, Biochemical Techniques. Separation and Analytical Techniques, Haematology, Physiology. The appendices in the last part are included to provide information important to the study of the above mentioned practicals as a whole.

This laboratory manual is the outcome of teaching in this area for past twenty-seven years to M.Sc. Zoology students, B.Sc. Honours students and students belonging to Biotechnology, Agriculture and allied fields. The idea behind this practical manual was thus to provide theoretical basis of the practical study items to be undertaken in the laboratory in a lucid manner. We shall be grateful if this book is accepted as a valued practical textbook by the students of Life Science at the undergraduate and postgraduate level of most Universities.

New Delhi   **Abhijit Dutta**

CONTENTS

ANIMAL CELL CULTURE

BIOCHEMICAL TECHNIQUES

SEPARATION AND ANALYTICAL TECHNIQUES

HAEMATOLOGY

PHYSIOLOGY

APPENDICES

INTRODUCTION

1

GENERAL INSTRUCTIONS ON USING THE LABORATORY

1. Laboratory experiments should be performed only when the instructor is present.
2. A laboratory apron and protective glasses or goggles for all laboratory work should be worn. Lab shoes should be worn (rather than sandals) and loose hair should be tied.
3. The working bench-top should be cleared of all unnecessary materials, such as books and clothing, before starting the work.
4. Chemical labels should be checked twice to make sure the correct for substance. Some chemical formula and names differ by only a letter or number. Attention should be paid to the hazard classification shown on the label.
5. Excess material should never be returned to its original container unless authorized by your instructor.
6. Unnecessary movement and talking inside the laboratory should be avoided.
7. Laboratory materials should never be tasted. Edibles should not be brought into the laboratory. In case it is required to smell something, it should be done by fanning some of the vapor toward one's nose. Never place your nose near the opening of the container.
8. Never look directly down into a test tube; view the contents from the side. Never point the open end of a test tube toward yourself or your neighbor.
9. Any laboratory accident, however small, should be reported immediately to the instructor.
10. Rinse the affected area with plenty of water in case of a chemical spill on the skin or clothing. If the eyes are affected washing with water must begin immediately and continued for 10 to 15 minutes or until professional assistance is obtained.
11. While discarding used chemicals, instructions provided should be carefully followed.
12. All equipments, chemicals, aprons, and protective glasses should be returned to their designated locations.
13. Ensure that gas lines and water faucets are shut off before leaving the laboratory.

PERSONAL PROTECTION

1. Laboratory coveralls, gowns or uniforms must be worn at all times for work in the laboratory.
2. Appropriate gloves must be worn for all procedures that may involve direct or accidental contact with blood, body fluids and other potentially infectious materials or infected animals. After use, gloves should be removed aseptically and hands must then be washed.
3. Personnel must wash their hands after handling infectious materials and animals, and before they leave the laboratory working areas.

4. Safety glasses, face shields (visors) or other protective devices must be worn when it is necessary to protect the eyes and face from splashes, impacting objects and sources of artificial ultraviolet radiation.
5. It is prohibited to wear protective laboratory clothing outside the laboratory, e.g. in canteens, coffee rooms, offices, libraries, staff rooms and toilets.
6. Open-toed footwear must not be worn in laboratories.
7. Eating, drinking, smoking, applying cosmetics and handling contact lenses is prohibited in the laboratory working areas.
8. Storing human foods or drinks anywhere in the laboratory working areas is prohibited.
9. Protective laboratory clothing that has been used in the laboratory must not be stored in the same lockers or cupboards as street clothing.

PROCEDURES

1. Pipetting by mouth must be strictly forbidden.
2. Materials must not be placed in the mouth. Labels must not be licked.
3. All technical procedures should be performed in a way that minimizes the formation of aerosols and droplets.
4. The use of hypodermic needles and syringes should be limited. They must not be used as substitutes for pipetting devices or for any purpose other than parenteral injection or aspiration of fluids from laboratory animals.
5. All spills, accidents and overt or potential exposures to infectious materials must be reported to the laboratory supervisor. A written record of such accidents and incidents should be maintained.
6. A written procedure for the clean-up of all spills must be developed and followed.
7. Contaminated liquids must be decontaminated (chemically or physically) before discharge to the sanitary sewer. An effluent treatment system may be required, depending on the risk assessment for the agent(s) being handled.
8. Written documents that are expected to be removed from the laboratory need to be protected from contamination while in the laboratory.

LABORATORY WORKING AREAS

1. The laboratory should be kept neat, clean and free of materials that are not pertinent to the work.
2. Working surfaces must be decontaminated after any spill of potentially dangerous material and at the end of the working day.
3. All contaminated materials, specimens and cultures must be decontaminated before disposal or cleaning for reuse.
4. Packing and transportation must follow applicable national and/or international regulations.
5. All windows to be fitted with arthopod-proof screens (firewire mesh).

2

BIOSAFETY

BIOSAFETY MANAGEMENT

1. It is the responsibility of the laboratory director (the person who has immediate responsibility for the laboratory) to ensure the development and adoption of a biosafety management plan and a safety or operations manual.
2. The laboratory supervisor (reporting to the laboratory director) should ensure that regular training in laboratory safety is provided.
3. Personnel should be advised of special hazards, and are required to read the safety or operations manual and follow standard practices and procedures. The laboratory supervisor should make sure that all personnel understand these. A copy of the safety or operations manual should be available in the laboratory.
4. There should be an arthropod and rodent control programme.
5. Appropriate medical evaluation, surveillance and treatment should be provided for all personnel in case of need, and adequate medical records should be maintained.

LABORATORY DESIGN AND FACILITIES

In designing a laboratory and assigning certain types of work to it, special attention should be paid to conditions that are known to pose safety problems. These include:
1. Formation of aerosols
2. Work with large volumes and/or high concentrations of microorganisms
3. Overcrowding and too much equipment
4. Infestation with rodents and arthropods
5. Unauthorized entrance
6. Workflow: use of specific samples and reagents.

Design Features

1. Ample space must be provided for the safe conduct of laboratory work and for cleaning and maintenance.
2. Walls, ceilings and floors should be smooth, easy to clean, impermeable to liquids and resistant to the chemicals and disinfectants normally used in the laboratory. Floors should be slip-resistant.
3. Bench tops should be impervious to water and resistant to disinfectants, acids, alkalis, organic solvents and moderate heat.

4. Illumination should be adequate for all activities. Undesirable reflections and glare should be avoided.

5. Laboratory furniture should be sturdy. Open spaces between and under benches, cabinets and equipment should be accessible for cleaning.

6. Storage space must be adequate to hold supplies for immediate use and thus prevent clutter on bench tops and in aisles. Additional long-term storage space, conveniently located outside the laboratory working areas, should also be provided.

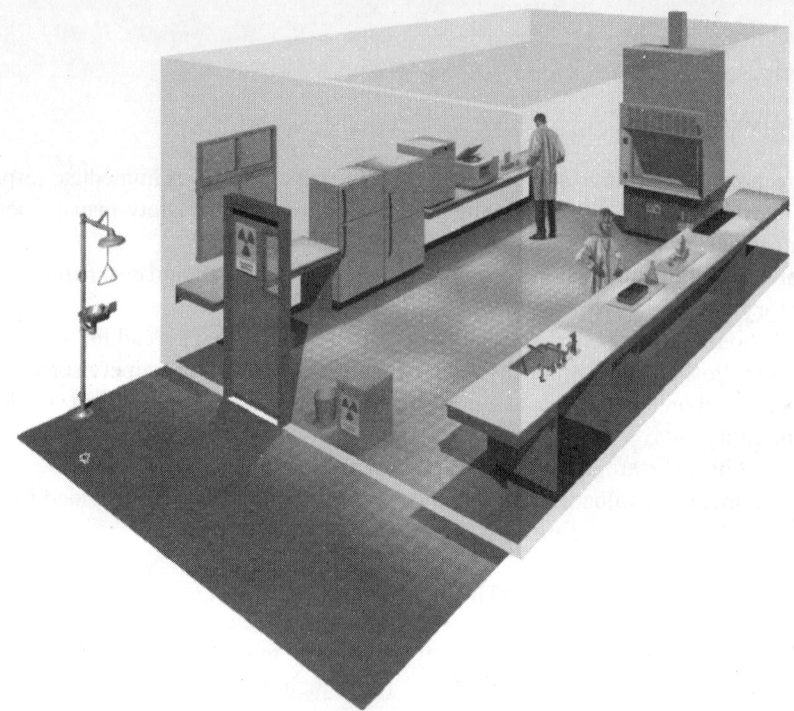

Fig. 1 A Typical Biosafety Laboratory

7. Space and facilities should be provided for safe handling and storage of solvents, radioactive materials, compressed and liquefied gases.

8. Facilities for storing outer garments and personal items should be provided outside the laboratory working areas.

9. Facilities for eating, drinking and resting should be provided outside the laboratory working area.

10. Wash basins, with running water if possible, should be provided in each laboratory room, preferably near the exit door.

11. Doors should have vision panels, appropriate fire ratings, and preferably be selfclosing.

12. Safety systems should cover fire, electrical emergencies, and emergency shower and eyewash facilities.

13. First-aid areas or rooms, suitably equipped and readily accessible, should be available.

14. In the planning of new facilities, consideration should be given to the provision of mechanical ventilation systems that provide an inward flow of air without recirculation.

15. A dependable supply of good quality water is essential. There should be no crossconnections between sources of laboratory and drinking-water supplies. An antibackflow device should be fitted to protect the public water system.
16. There should be a reliable and adequate supply of electricity and emergency lighting to permit safe exit. A stand-by generator is desirable for the support of essential equipment, such as incubators, biological safety cabinets, freezers, etc., and for the ventilation of animal cages.
17. There should be a reliable and adequate supply of gas. Good maintenance of the installation is mandatory.
18. Laboratories and animal houses are occasionally the targets of vandals. Physical and fire security must be considered. Strong doors, screened windows and restricted issue of keys are compulsory. Other measures should be considered and applied, as appropriate, to augment security.

ESSENTIAL BIOSAFETY EQUIPMENT

1. Pipetting aids.
2. Biological safety cabinets; Equipment such as autoclaves and biological safety cabinets must be validated with appropriate methods before being taken into use. Recertification should take place at regular intervals, according to the manufacturer's instructions
3. Plastic disposable transfer loops. Alternatively, electric transfer loop incinerators may be used inside the biological safety cabinet to reduce aerosol production.
4. Screw-capped tubes and bottles.
5. Autoclaves or other appropriate means to decontaminate infectious materials.
6. Plastic disposable Pasteur pipettes.

Waste Handling

Waste is anything that is to be discarded. In laboratories, decontamination of wastes and their ultimate disposal are closely interrelated. In terms of daily use, few if any contaminated materials will require actual removal from the laboratory or destruction. Most glassware, instruments and laboratory clothing will be reused or recycled. The overriding principle is that all infectious materials should be decontaminated, autoclaved or incinerated within the laboratory.

Decontamination

Steam autoclaving is the preferred method for all decontamination processes. Materials for decontamination and disposal should be placed in containers, e.g. autoclavable plastic bags that are colour-coded according to whether the contents are to be autoclaved and/or incinerated. Alternative methods may be envisaged only if they remove and/or kill microorganisms.

Handling and Disposal Procedures for Contaminated Materials and Wastes

Identification and separation system for infectious materials and their containers should be adopted. National and international regulations must be followed. Categories should include:

1. Non-contaminated (non-infectious) waste that can be reused or recycled or disposed of as general, "household" waste
2. Contaminated (infectious) "sharps" – hypodermic needles, scalpels, knives and broken glass; these should always be collected in puncture-proof containers fitted with covers and treated as infectious containing disinfecting solution.

3. Contaminated material for decontamination by autoclaving and thereafter washing and reuse or recycling
4. Contaminated material for autoclaving and disposal
5. Contaminated material for direct incineration.

LABORATORY DESIGN AND FACILITIES

The laboratory design and facilities for basic laboratories apply except where modified as follows:

1. The laboratory must be separated from the unrestricted area within the building.
2. Surfaces of walls, floors and ceilings should be water-resistant and easy to clean. Openings through these surfaces (e.g. for service pipes) should be sealed to facilitate decontamination of the room(s).
3. Windows must be closed, sealed and break-resistant.
4. There must be a controlled ventilation system that maintains a directional airflow into the laboratory room.
5. An autoclave for decontamination of contaminated waste material should be available.

Safe Handling of Specimens in the Laboratory

Improper collection, transport and handling of specimens in the laboratory carry a risk of infection to the personnel involved.

Specimen Containers

Specimen containers may be of glass, but preferably of plastic material. They should be robust and should not leak when the cap or stopper is correctly applied. No material should remain on the outside of the container. Containers should be correctly labelled to facilitate identification. Specimen request or specification forms should not be wrapped around the containers but placed in separate, preferably waterproof envelopes.

Transport of Specimens within the Facility

To avoid accidental leakage or spillage, secondary containers, such as boxes, should be used, fitted with racks so that the specimen containers remain upright. The secondary containers may be of metal or plastic, should be autoclavable or resistant to the action of chemical disinfectants, and the seal should preferably have a gasket. They should be regularly decontaminated.

Receipt of Specimens

Laboratories that receive large numbers of specimens should designate a particular room or area for this purpose.

Opening Packages

Personnel who receive and unpack specimens should be aware of the potential health hazards involved, and should be trained to adopt standard precautions, particularly when dealing with broken or leaking containers. Primary specimen containers should be opened in a biological safety cabinet. Disinfectants should be available.

USE OF PIPETTES AND PIPETTING AIDS

1. A pipetting aid must always be used. Pipetting by mouth must be prohibited.
2. All pipettes should have cotton plugs to reduce contamination of pipetting devices.
3. Air should never be blown through liquid, containing infectious agents.
4. Infectious materials should not be mixed by alternate suction and expulsion through a pipette.
5. Liquids should not be forcibly expelled from pipettes.
6. Mark-to-mark pipettes are preferable to other types as they do not require expulsion of the last drop.
7. Contaminated pipettes should be completely submerged in a suitable disinfectant contained in an unbreakable container. They should be left in the disinfectant for the appropriate length of time before disposal.
8. A discard container for pipettes should be placed within the biological safety cabinet, not outside it.
9. Syringes fitted with hypodermic needles must not be used for pipetting.
10. Devices for opening septum-capped bottles that allow pipettes to be used and avoid the use of hypodermic needles and syringes should be used.
11. To avoid dispersion of infectious material dropped from a pipette, an absorbent material should be placed on the working surface; this should be disposed of as infectious waste after use.

DISPERSAL OF INFECTIOUS MATERIALS

1. In order to avoid premature shedding of loads, microbiological transfer loops should have a diameter of 2–3 mm and be completely closed. The shanks should be not more than 6 cm in length to minimize vibration.
2. The risk of spatter of infectious material in an open Bunsen burner flame should be avoided by using an enclosed electric microincinerator to sterilize transfer loops. Disposable transfer loops, which do not need to be resterilized, are preferable.
3. Care should be taken when drying sputum samples, to avoid creating aerosols.
4. Discarded specimens and cultures for autoclaving and/or disposal should be placed in leakproof containers, e.g. laboratory discard bags. Tops should be secured (e.g. with autoclave tape) prior to disposal into waste containers.
5. Working areas must be decontaminated with a suitable disinfectant at the end of each work period.

USE OF BIOLOGICAL SAFETY CABINETS

1. The advantages and limitations of biological safety cabinets should be explained to all potential users. Written protocols or operation manuals should be issued to students. In particular, it must be made clear that the cabinet will not protect the operator from spillage, breakage or poor technique.
2. The cabinet must not be used unless it is working properly.
3. The glass viewing panel must not be opened when the cabinet is in use.
4. Apparatus and materials in the cabinet must be kept to a minimum. Air circulation at the rear plenum must not be blocked.
5. Bunsen burners must not be used in the cabinet. The heat produced will distort the airflow and may damage the filters. An electric microincinerator is permissible but sterile disposable transfer loops are better.

6. All work must be carried out in the middle or rear part of the working surface and be visible through the viewing panel.
7. Traffic behind the operator should be minimized.
8. The operator should not disturb the airflow by repeated removal and reintroduction of his or her arms.
9. Air grills must not be blocked with notes, pipettes or other materials, as this will disrupt the airflow causing potential contamination of the material and exposure of the operator.
10. The surface of the biological safety cabinet should be wiped using an appropriate disinfectant after work is completed and at the end of the day.
11. The cabinet fan should be run for at least 5 min before beginning work and after completion of work in the cabinet.
12. Paperwork should never be placed inside biological safety cabinets.

AVOIDING INGESTION OF INFECTIOUS MATERIALS AND CONTACT WITH SKIN AND EYES

1. Large particles and droplets (> 5 μm in diameter) released during microbiological manipulations settle rapidly on bench surfaces and on the hands of the operator. Disposable gloves should be worn. Laboratory workers should avoid touching their mouth, eyes and face during lab works.
2. Food and drink must not be consumed or stored in the laboratory.
3. No article should be placed into the mouth - e.g. pens, pencils, chewing gums etc., once inside the laboratory.
4. Cosmetics should not be applied in the laboratory.
5. The face, eyes and mouth should be shielded or otherwise protected during any operation that may result in the splashing of potentially infectious materials.

PREVENTION OF INJECTION WITH INFECTIOUS MATERIALS

1. Accidental inoculation resulting from injury with broken or chipped glassware can be avoided through careful practices and procedures. Glassware should be replaced with plastic ware whenever possible.
2. Accidental injection may result from sharps injuries e.g. with hypodermic needles (needle-sticks), glass Pasteur pipettes, or broken glass.
3. Needle-stick injuries can be reduced by:

    (a) Minimizing the use of syringes and needles (e.g. simple devices are available for opening septum-stoppered bottles so that pipettes can be used instead of syringes and needles; or
    (b) Using engineered harp safety devices when syringes and needles are necessary.

4. Needles should never be recapped. Disposable articles should be discarded into puncture-proof/puncture-resistant containers fitted with covers.
5. Plastic Pasteur pipettes should replace those made of glass.

SEPARATION OF SERUM

1. Only properly trained staff should be employed for this work.
2. Gloves, glasses should be worn.

3. Splashes and aerosols can only be avoided or minimized by good laboratory technique. Blood and serum should be pipetted carefully, not poured. Pipetting by mouth must be forbidden.

4. After use, pipettes should be completely submerged in suitable disinfectant. They should remain in the disinfectant for the appropriate time before disposal or washing and sterilization for reuse.

5. Discarded specimen tubes containing blood clots, etc. (with caps replaced) should be placed in suitable leakproof containers for autoclaving and/or incineration.

6. Suitable disinfectants should be available for clean-up of splashes and spillages.

USE OF CENTRIFUGES

1. Satisfactory mechanical performance is a prerequisite for microbiological safety in the use of laboratory centrifuges.

2. Centrifuges should be operated according to the manufacturer's instructions.

3. Centrifuges should be placed at such a level that workers are able to look inside the bowl to place trunnions and buckets correctly.

4. Centrifuge tubes and specimen containers for use in the centrifuge should be made of thick-walled glass or preferably of plastic and should be inspected for defects before use.

5. Tubes and specimen containers should always be securely capped (screw-capped if possible) for centrifugation.

6. The buckets must be loaded, equilibrated, sealed and opened in a biological safety cabinet.

7. Buckets and trunnions should be paired by weight and, with tubes in place, correctly balanced.

8. The amount of space that should be left between the level of the fluid and the rim of the centrifuge tube should be as per the specifications in the manufacturer's instructions.

9. Distilled water or alcohol (propanol, 70%) should be used for balancing empty buckets. Saline or hypochlorite solutions should not be used as they corrode metals.

10. Sealable centrifuge buckets (safety cups) must be used for microorganisms

11. When using angle-head centrifuge rotors, care must be taken to ensure that the tube is not overloaded as it might leak.

12. The interior of the centrifuge bowl should be inspected daily for staining or soiling at the level of the rotor. If staining or soiling is evident then the centrifugation protocols should be re-evaluated.

13. Centrifuge rotors and buckets should be inspected daily for signs of corrosion and for hair-line cracks.

14. Buckets, rotors and centrifuge bowls should be decontaminated after each use.

15. After use, buckets should be stored in an inverted position to drain the balancing fluid.

USE OF HOMOGENISERS, SHAKERS, BLENDERS AND SONICATORS

1. Domestic (kitchen) homogenisers should not be used in laboratories as they may leak or release aerosols. Laboratory blenders and stomachers are safer.

2. Caps and cups or bottles should be in good condition and free from flaws or distortion. Caps should be well-fitting and gaskets should be in good condition.

3. Pressure builds up in the vessel during the operation of homogenisers, shakers and sonicators. Aerosols containing infectious materials may escape from between the cap and the vessel. Plastic, in particular, polytetrafluoroethylene (PTFE) vessels are recommended because glass may break, releasing infectious material and possibly wounding the operator.

4. When in use, homogenisers, shakers and sonicators should be covered by a strong transparent plastic casing. This should be disinfected after use. Where possible, these machines should be operated, under their plastic covers, in a biological safety cabinet.
5. At the end of the operation the containers should be opened in a biological safety cabinet.
6. Ear muffs should be provided for people using sonicators.

USE OF TISSUE GRINDERS

1. Glass grinders should be held in absorbent material in a gloved hand. Plastic grinders are safer.
2. Tissue grinders should be operated and opened in a biological safety cabinet.

CARE AND USE OF REFRIGERATORS AND FREEZERS

1. Refrigerators, deep-freezers and solid carbon dioxide (dry-ice) chests should be defrosted and cleaned periodically, and any ampoules, tubes, etc. that have broken during storage be removed. Face protection and heavy duty rubber gloves should be worn during cleaning. After cleaning, the inner surfaces of the cabinet should be disinfected.
2. All containers stored in refrigerators, etc. should be clearly labelled with the scientific name of the contents, the date stored and the name of the individual who stored them. Unlabelled and obsolete materials should be autoclaved and discarded.
3. An inventory of the freezer's contents must be maintained.
4. Flammable solutions must not be stored in a refrigerator unless it is explosion proof. Notices to this effect should be placed on refrigerator doors.

OPENING OF AMPOULES CONTAINING LYOPHILIZED INFECTIOUS MATERIALS

Care should be taken when ampoules of freeze-dried materials are opened, as the contents may be under reduced pressure and the sudden inrush of air may disperse some of the materials into the atmosphere. Ampoules should always be opened in a biological safety cabinet. The following procedures are recommended for opening ampoules.

1. First decontaminate the outer surface of the ampoule.
2. Make a file mark on the tube near to the middle of the cotton or cellulose plug, if present.
3. Hold the ampoule in alcohol-soaked cotton to protect hands before breaking it at a file scratch.
4. Remove the top gently and treat as contaminated material.
5. If the plug is still above the contents of the ampoule, remove it with sterile forceps.
6. Add liquid for resuspension slowly to the ampoule to avoid frothing.

STORAGE OF AMPOULES CONTAINING INFECTIOUS MATERIALS

Ampoules containing infectious materials should never be immersed in liquid nitrogen because cracked or imperfectly sealed ampoules may break or explode on removal. If very low temperatures are required, ampoules should be stored only in the gaseous phase above the liquid nitrogen. Otherwise, infectious materials should be stored in mechanical deep-freeze cabinets or on dry ice. Laboratory workers should wear eye and hand protection when removing ampoules from cold storage. The outer surfaces of ampoules stored in these ways should be disinfected when the ampoules are removed from storage.

STANDARD PRECAUTIONS WITH BLOOD AND OTHER BODY FLUIDS, TISSUES AND EXCRETA

Collection, Labelling and Transport of Specimens

1. Standard precautions (2) should always be followed; gloves should be worn for all procedures.
2. Blood should be collected from patients and animals by trained staff.
3. For phlebotomies, conventional needle and syringe systems should be replaced by single-use safety vacuum devices that allow the collection of blood directly into stoppered transport and/or culture tubes, automatically disabling the needle after use.
4. The tubes should be placed in adequate containers for transport to the laboratory within the laboratory facility and should be placed in separate waterproof bags or envelopes.

Opening Specimen Tubes and Sampling Contents

1. Specimen tubes should be opened in a biological safety cabinet.
2. Gloves must be worn. Eye and mucous membrane protection is also recommended (goggles or face shields).
3. Protective clothing should be supplemented with a plastic apron.
4. The stopper should be grasped through a piece of paper or gauze to prevent splashing.

Glass and "Sharps"

1. Plastics should replace glass wherever possible. Only laboratory grade (borosilicate) glass should be used, and any article that is chipped or cracked should be discarded.
2. Hypodermic needles must not be used as pipettes.

Films and smears for microscopy

Fixing and staining of blood, sputum and faecal samples for microscopy do not necessarily kill all organisms or viruses on the smears. These items should be handled with forceps, stored appropriately, and decontaminated and/or autoclaved before disposal.

Automated Equipment (Sonicators, Vortex Mixers)

1. Equipment should be of the closed type to avoid dispersion of droplets and aerosols.
2. Effluents should be collected in closed vessels for further autoclaving and/or disposal.
3. Equipment should be disinfected at the end of each session, following manufacturers' instructions.

Tissues

1. Formalin fixatives should be used.
2. Frozen sectioning should be avoided. When necessary, the cryostat should be shielded and the operator should wear a safety face shield. For decontamination, the temperature of the instrument should be raised to at least 20 °C.

Decontamination

Hypochlorites and high-level disinfectants are recommended for decontamination. Freshly prepared hypochlorite solutions should contain available chlorine at 1 g/l for general use and 5 g/l for blood spillages. Glutaraldehyde may be used for decontaminating surfaces

Emergency equipment

The following emergency equipment must be available:

1. First-aid kit, including universal and special antidotes
2. Appropriate fire extinguishers, fire blankets

The following are also suggested but may be varied according to local circumstances:

1. Full protective clothing (one-piece coveralls, gloves and head covering – for incidents involving microorganisms in Risk Groups 3 and 4)
2. Full-face respirators with appropriate chemical and particulate filter canisters
3. Room disinfection apparatus, e.g. sprays and formaldehyde vaporizers
4. Stretcher
5. Tools, e.g. hammers, axes, spanners, screwdrivers, ladders, ropes
6. Hazard area demarcation equipment and notices

CLEANING LABORATORY MATERIALS

Cleaning is the removal of dirt, organic matter and stains. Cleaning includes brushing, vacuuming, dry dusting, washing or damp mopping with water containing a soap or detergent. Dirt, soil and organic matter can shield microorganisms and can interfere with the killing action of decontaminants (antiseptics, chemical germicides and disinfectants). Precleaning is essential to achieve proper disinfection or sterilization. Many germicidal products claim activity only on precleaned items. Precleaning must be carried out with care to avoid exposure to infectious agents. Materials chemically compatible with the germicides to be applied later must be used. It is quite common to use the same chemical germicide for precleaning and disinfection.

AUTOCLAVING

Saturated steam under pressure (autoclaving) is the most effective and reliable means of sterilizing laboratory materials. For most purposes, the following cycles will ensure sterilization of correctly loaded autoclaves:

1. 3 min holding time at 134 °C
2. 10 min holding time at 126 °C
3. 15 min holding time at 121 °C
4. 25 min holding time at 115 °C

Source:
- *Recombinant DNA Safety Guidelines and Regulations, Dept. of Biotechnology, Ministry of Science and Technology, New Delhi (Jan. 1990)*
- *Laboratory Biosaafety Manual, WHO, 2004*

3

SOME USEFUL TIPS REGARDING WEIGHTS, MEASUREMENTS, SOLUTION PREPARATION AND CALCULATIONS

ATOMIC WEIGHT

atomic weight, mean (weighted average) of the masses of all the naturally occurring isotopes of a chemical element, as contrasted with atomic mass, which is the mass of any individual isotope. Although the first atomic weights were calculated at the beginning of the 19th cent., it was not until the discovery of isotopes by F. Soddy (c.1913) that the atomic mass of many individual isotopes was determined, leading eventually to the adoption of the atomic mass unit as the standard unit of atomic weight.

GRAM-ATOMIC WEIGHT

Gram-atomic weight, amount of an atomic substance whose weight, in grams, is numerically equal to the atomic weight of that substance. For example, 1 gram-atomic weight of atomic oxygen, O (atomic weight approximately 16), is 16 grams.

ATOMIC WEIGHT AND NUMBER

The atomic number of an atom is simply the number of protons in its nucleus. The atomic weight of an atom is given in most cases by the mass number of the atom, equal to the total number of protons and neutrons combined. An atom may be conveniently symbolized by its chemical symbol with the atomic number and mass number written as subscript and superscript, respectively. For example, the symbol for uranium is U (atomic number 92); the isotopes of uranium with atomic weights 235 and 238 are indicated by $^{235}_{92}U$ and $^{238}_{92}U$.

MOLECULAR WEIGHT

The sum of the atomic weights of all the atoms in a molecule. Also called *formula weight*, molecular weight, or weight of a molecule of a substance expressed in atomic mass units (amu). The molecular weight may be calculated from the molecular formula of the substance; it is the sum of the atomic weights of the atoms making up the molecule. For example, water has the molecular formula H_2O, indicating that there are two atoms of hydrogen and one atom of oxygen in a molecule of water. Rounded to three decimal places, the atomic weight of hydrogen is 1.008 amu and that of oxygen is 15.999 amu. The molecular weight of water is thus $(2 \times 1.008) + (1 \times 15.999) = 2.016 + 15.999 = 18.015$ amu. Since atomic weights are average values, molecular weights are also average values. On the average, a molecule of ordinary water weighs 18.015 amu. Both hydrogen and oxygen are made up of several isotopes. One isotope of hydrogen is deuterium, or heavy hydrogen. Atoms of deuterium are about twice as massive as the average for all hydrogen atoms in ordinary water. Therefore water that contains only atoms of deuterium, called heavy water, has a higher molecular weight than ordinary water. Some substances, especially ionic compounds such as common salt, are not made up of molecules and thus have neither a molecular formula nor a molecular weight.

Molecular weights of substances may be determined experimentally in various ways, the method employed usually depending on the state (solid, liquid, or gas) of the substance. Methods for determining the molecular weights of gaseous substances are based on Avogadro's law, which states that under given conditions of temperature and pressure a given volume of any gas contains a specific number of molecules of the gas; thus a comparison of the weights of equal volumes of different gases under the same conditions of temperature and pressure is equivalent to a direct comparison of the weights of molecules of the gases. The molecular weights of substances that are not normally gaseous and do not evaporate without decomposition are sometimes determined from their effects on the melting point, boiling point, vapor pressure, or osmotic pressure of some solvent (see colligative properties). However, if the substance ionizes or does not completely separate into molecules, the molecular weight so determined will be erroneous. Highly accurate molecular weights are sometimes determined by using the mass spectrograph.

Some substances, e.g., proteins, viruses, and certain synthetic polymers, have very high molecular weights. These molecular weights may be determined by measurement of sedimentation rate in an ultracentrifuge, by light-scattering photometry, or by other methods. The methods may give different results, since usually the molecules of a substance such as a polymer do not all have exactly the same molecular weight. These methods determine an average molecular weight for the molecules in the sample. The number-average molecular weight determined by the ultracentrifuge method gives a value that is equal to the weight of the sample divided by the number of molecules in the sample. This number-average molecular weight can also be determined by other methods based on measurement of colligative properties. The light-scattering method determines what is called the weight-average molecular weight. Although this may be the same value as the number-average molecular weight if all the molecules have nearly the same weight, it will be higher if some of the molecules are heavier than others.

Calculation and preparation of solutions

Abbreviations
    l = liter
    g = grams
    wt = weight
    MW = molecular weight
    FW = formula weight

METRIC SYSTEM
    kilo – 1000 (k)
    centi – 1/100 or 10^2
    milli – 1/1000 or 10^3 (m)
    micro – 1/1,000,000 or 10^6 (u)
    nano – 1/1,000,000,000 or 10^9 (n)
    pico – 1/1,000,000,000,000 or 10^{12} (p)

Hint: To convert one unit to another (except for centi), move the decimal point three spaces

```
0.3 mg =  300 ug
0.3 ml =  300 ul
```

MOLAR SOLUTIONS

A. Mole (mol)

One mole of atom is the number of atoms (6.02×10^{23}) in exactly 12 g (12 grams) of ^{12}C. The weight in grams is equal to the numerical value of the atomic weight or molecular weight. One mole of calcium weighs 40.08 g and contains 6.02×10^{23} atoms.

$$mole = weight\ (g)/molecular\ weight\ (g/mol)$$

B. Molecular weight and gram molecular weight (MW)

The molecular weight of a compound is the sum of the atomic weight of all the atoms that make up one molecule of the compound. An amount in gram molecular weight of any compound contains 1 mole (6.02×1023 molecules) of the compound. The molecular weight expressed in grams is termed the gram molecular weight but is often shortened to just "molecular weight".

C. Formula weight (FW)

This unit is used in place of molecular weight to designate the weight of a compound that does not exist as a discrete molecule. It is the sum of the atomic weights of all elements that comprise the chemical formula of the compound.

D. Molarity (M)

Molarity is expressed as moles per liter. Thus, a 2 molar solution has 2 moles of a substance in a total final volume of 1 liter. Use the periodic table in a chemistry text or the Handbook of Chemistry and Physics to look up the molecular weights of individual atoms in a compound.

$$M = moles/liter$$

and

$$M = wt\ (g)/\ mol\ wt \times liters$$

How many grams are in a mole?

This is different for different elements or compounds. This is given by the atomic mass of an element, or by the Molecular Weight (M.W.) (may also be called the Formula Weight (F.W.); in most cases these are the same) of a compound, which is the sum of atomic masses of the atoms found in one formula unit of a compound. For example, the formula weight of water ($H2O$) is two times the atomic weight of hydrogen plus one times the atomic weight of oxygen. Numerically, this is

$(2*1.00797) + (1*15.9994) = 2.01594+15.9994 = 18.01534.$

Another example:

The formula weight of KCl = 39.1 + 35.45 = 74.55. To make a M solution of KCl, you need 74.55 grams per liter. **Note:** the above means 74.55 grams per liter of final solution. This is not the same as 74.55 grams added to 1 liter of water. When making the solution, the solute should be added to a smaller volume of water first, and then the final volume adjusted to 1 liter.

What is Normality?

Normality (N) is often used for acids and bases. A very simple definition is that

Normality = Molarity x Z, where Z represents the number of protons (H+) donated by 1 molecule of an acid, or the number of hydroxide ions (OH-) donated by 1 molecule of a base. Z is typically 1, 2, or 3.

Examples:

$$\text{Hydrochloric acid, HCl, } Z = 1$$
$$\text{Calcium Hydroxide, Ca(OH)2, } Z = 2$$
$$\text{Phosphoric acid, H3PO4, } Z = 3$$

Thus, to make a 1 N solution of Ca(OH)2, (F.W. = 74.09) you need: 1/2 mole per liter or 37.05 grams/liter.

Practical Hints

The molecular weight or formula weight of a chemical is usually printed somewhere on the label or it is possible to calculate the molecular weight from the formula using a periodic table. To make a one molar solution, dissolve 1 mole (in grams) in a final volume of 1 liter of solution. The abbreviation Q.S. (quantity sufficient) is often used to indicate that liquid is added to a given final volume (e.g., Q.S. 1 liter).

Dilutions

A. Stock solutions are often prepared and must be diluted to final concentration. An equation relating the various parameters is as follows:

$$V_1 = \text{ volume of stock reagent}$$
$$C_1 = \text{ concentration of the stock reagent}$$
$$V_2 = \text{ final volume needed}$$
$$C_2 = \text{ final concentration needed}$$
$$C_1 / C_2 = V_2 / V_1$$

Solve for the desired parameter. V1 is usually the unknown (the volume of stock needed to make a final concentration. Thus:

$$V_1 = [V_2 \times C_2]/C_1$$

What is a 1% (vol/vol) solution?

This means that the specified liquid chemical is 1% of the total solution volume. For example, a 1% glycerol solution would be 1 ml glycerol in a final volume of 100 ml.

What is a 1% (wt/vol) solution?

When the chemical is solid, a 1% solution means 1 gram per 100 ml. For example, a 1% SDS solution

Hint: To convert one unit to another (except for centi), move the decimal point three spaces

0.3 mg = 300 ug
0.3 ml = 300 ul

MOLAR SOLUTIONS

A. Mole (mol)

One mole of atom is the number of atoms (6.02×10^{23}) in exactly 12 g (12 grams) of ^{12}C. The weight in grams is equal to the numerical value of the atomic weight or molecular weight. One mole of calcium weighs 40.08 g and contains 6.02×10^{23} atoms.

$$\text{mole} = \text{weight (g)}/\text{molecular weight (g/mol)}$$

B. Molecular weight and gram molecular weight (MW)

The molecular weight of a compound is the sum of the atomic weight of all the atoms that make up one molecule of the compound. An amount in gram molecular weight of any compound contains 1 mole (6.02 X 1023 molecules) of the compound. The molecular weight expressed in grams is termed the gram molecular weight but is often shortened to just "molecular weight".

C. Formula weight (FW)

This unit is used in place of molecular weight to designate the weight of a compound that does not exist as a discrete molecule. It is the sum of the atomic weights of all elements that comprise the chemical formula of the compound.

D. Molarity (M)

Molarity is expressed as moles per liter. Thus, a 2 molar solution has 2 moles of a substance in a total final volume of 1 liter. Use the periodic table in a chemistry text or the Handbook of Chemistry and Physics to look up the molecular weights of individual atoms in a compound.

$$M = \text{moles/liter}$$

and

$$M = \text{wt (g)}/\text{ mol wt} \times \text{liters}$$

How many grams are in a mole?

This is different for different elements or compounds. This is given by the atomic mass of an element, or by the Molecular Weight (M.W.) (may also be called the Formula Weight (F.W.); in most cases these are the same) of a compound, which is the sum of atomic masses of the atoms found in one formula unit of a compound. For example, the formula weight of water (H2O) is two times the atomic weight of hydrogen plus one times the atomic weight of oxygen. Numerically, this is

(2*1.00797) + (1*15.9994) = 2.01594+15.9994 = 18.01534.

Another example:

The formula weight of KCl = 39.1 + 35.45 = 74.55. To make a M solution of KCl, you need 74.55 grams per liter. **Note:** the above means 74.55 grams per liter of final solution. This is not the same as 74.55 grams added to 1 liter of water. When making the solution, the solute should be added to a smaller volume of water first, and then the final volume adjusted to 1 liter.

What is Normality?

Normality (N) is often used for acids and bases. A very simple definition is that

Normality = Molarity x Z, where Z represents the number of protons (H+) donated by 1 molecule of an acid, or the number of hydroxide ions (OH-) donated by 1 molecule of a base. Z is typically 1, 2, or 3.

Examples:

Hydrochloric acid, HCl, Z = 1
Calcium Hydroxide, Ca(OH)2, Z = 2
Phosphoric acid, H3PO4, Z = 3

Thus, to make a 1 N solution of Ca(OH)2, (F.W. = 74.09) you need: 1/2 mole per liter or 37.05 grams/liter.

Practical Hints

The molecular weight or formula weight of a chemical is usually printed somewhere on the label or it is possible to calculate the molecular weight from the formula using a periodic table. To make a one molar solution, dissolve 1 mole (in grams) in a final volume of 1 liter of solution. The abbreviation Q.S. (quantity sufficient) is often used to indicate that liquid is added to a given final volume (e.g., Q.S. 1 liter).

Dilutions

A. Stock solutions are often prepared and must be diluted to final concentration. An equation relating the various parameters is as follows:

V_1 = volume of stock reagent
C_1 = concentration of the stock reagent
V_2 = final volume needed
C_2 = final concentration needed
$C_1 / C_2 = V_2 / V_1$

Solve for the desired parameter. V1 is usually the unknown (the volume of stock needed to make a final concentration. Thus:

$$V_1 = [V_2 \times C_2]/C_1$$

What is a 1% (vol/vol) solution?

This means that the specified liquid chemical is 1% of the total solution volume. For example, a 1% glycerol solution would be 1 ml glycerol in a final volume of 100 ml.

What is a 1% (wt/vol) solution?

When the chemical is solid, a 1% solution means 1 gram per 100 ml. For example, a 1% SDS solution weight approximately 16), is 16 grams. See gram-molecular weight .

Quick Reference to Selected Equations

Equations for calculating amounts and concentrations of reagents

Mole (mol)

1 mole of any substance contains 6.02×10^{23} units. The number of moles in a given weight of substance can be calculated using equation 1, and the weight of a given mount of moles can be calculated using equation-2

$$\text{mol} = \frac{\text{wt (g)}}{\text{mol wt}} \tag{1}$$

$$\text{wt(g)} = \text{mol} \times \text{mol wt (g/mol)} \tag{2}$$

Equivalent Weight (equiv wt)

This is the weight of an acid or a base containing 1 mole of replaceable H^+ or OH^-. Gram equivalent weight is used when equivalent weight is expressed in grams, as calculated in equation -3

$$\text{equiv wt} = \text{mol wt}/n \tag{3}$$

Equivalent (equiv)

One equivalent of a compound contain 1 g-equiv wt of the compound, as calculated in equation -4

$$\text{equiv} = \frac{\text{wt (g)}}{\text{equiv wt}} \text{ or equiv} = \text{mol} \times n \tag{4}$$

Molarity (M)

This defines the number of moles of a substance per liter (L) of solution, as calculated in equation -5.

$$M = \frac{\text{mol}}{L} \text{ or } M = \text{mol} \times n \tag{5}$$

Normality (N)

This defines the number of equiv of a substance per liter of solution, as calculated in equation 6 and 7.

$$N = \frac{\text{equiv}}{L} \text{ or } N = \frac{\text{wt}}{\text{mol wt} \times L} \tag{6}$$

$$N = n M \tag{7}$$

CALCULATING CONCENTRATIONS BASED ON WEIGHT

Percent (Weight/Weight) or %(w/w)

The percent weight of a pure analyte in a sample is calculated using equation 8 (e.g the weight in gm of an analyte in a 100g sample).

$$\% \, (w/w) = \frac{g \text{ of analyte in sample}}{g \text{ of sample}} \times 100\% \tag{8}$$

Percent (Weight/Volume) or %(w/v)

The percent weight of a pure analyte in a solution is calculated using equation 9 (e.g. the weight in g of an analyte in 100ml sample)

$$\% \, (w/v) = \frac{g \text{ of analyte in sample}}{mL \text{ of sample}} \times 100\% \tag{9}$$

Parts Per Million (Weight/Weight), or ppm(w/w)

The grams of a pure analyte in 10^6 g of sample is calculated using equation 10.

$$ppm \, (w/v) = \frac{g \text{ of analyte in sample}}{mL \text{ of sample}} \times 10^6 \, ppm \tag{10}$$

Part Per Million (Weight/Volume) or ppm (w/v)

The grams of pure analyte in 10^6 mL of sample is calculated using equation 11.

$$ppm \, (w/v) = \frac{g \text{ of analyte in sample}}{mL \text{ of sample}} \times 10^6 \tag{11}$$

Density (p) and Specific Gravity (sp gr)

Density is the mass per volume ratio of a substance and is calculated using equation 12.

$$\rho = \frac{wt}{vol} \tag{12}$$

Specific gravity is the density of a fluid or other material relative to that of H_2O as expressed in equation 13.

$$sp \, gr = \frac{\rho \, sample}{\rho \, H_2o} \quad \rho \, sample \text{ at } 4° \text{ C} \tag{13}$$

Equations for Calculating the Weight or Volume of a Reagent needed to prepare a given solution

To calculate the number of gram of a substance to create a pre-assigned molar concentration use equation-14

$$wt\ (g) = mol\ wt\ (g) \times M \times L \qquad (14)$$

When correcting for % purity use equation -15

$$wt\ (g) = mol\ wt\ (g) \times M \times L \times \frac{100}{purity\ of\ reagent(\%)} \qquad (15)$$

For % and ppm concentration use equation 16 and 17, respectively

$$wt\ (g) = \frac{\%\ (w/v) \times mL\ of\ solution}{100\%}$$

$$wt\ (g) = \frac{\%\ (w/v) \times mL\ of\ solution}{10^{-6}\ ppm} \qquad (16\ \&\ 17)$$

Dilution Factor in Concentration and Volume Calculations

To calculate the volume of a stock reagent necessary to add to a known volume of diluent to yield a desired concentration use equation -18

$$V_1 = \frac{C_2 V_2}{C_1} \qquad (18)$$

where V_1 is the volume of the stock reagent which, when diluted to a final volume V_2, yields a desired concentration C_2; C_1 is the concentration of the stock reagent. The units of C_1 and C_2 must be identical, and those of V_1 and V_2 must also be identical.

Reporting Analysis Results Based On Moisture

To convert contents (%) Y from oven-dried (OD) to an as-received (AR) basis (equation 19), or vice versa (equation 20), use the following formulas:

$$\%\ Y_{OD} = \frac{\%\ Y_{AR} \times 100}{(100 - \%\ loss_{OD})} \qquad (19)$$

$$\%\ Y_{AR} = \frac{\%\ Y_{AD}\ (100 - \%\ loss_{OD})}{100} \qquad (20)$$

Where: $\%loss_{OD} = 100 \times$ wet weight - dry weight
To convert contents from an AR basis to an arbitrary moisture AM basis use equation 21:

$$\%\ Y = \frac{\%\ Y_{AR}\ (100 - arbitrary\ moisture\ \%)}{100 - \%\ mositure_{AR}} \qquad (21)$$

To weigh out a sample on an arbitrary moisture basis use equation 22:

$$sample\ weight = \frac{\%\ dry\ matter\ _{AM}}{\%\ dry\ matter_{AR}} \times required\ sample\ wt. \qquad (22)$$

SOME BASIC CONVERSIONS AND CALCULATIONS

1. Conversions

1 gram (g)	=	1000 milligrams (mg)
1 milligram (mg)	=	1000 microgram (ug)
1 microgram (ug)	=	1000 nanogram (ng)
1 nanogram (ng)	=	1000 picogram (pg)
1 gram	=	10e12 picograms
1 mole	=	1000 millimoles (mmole)
1 millimole (mmole)	=	1000 micromoles (umole)
1 micromole (umole)	=	1000 nanomoles (nmole)
1 nanomole (nmole)	=	1000 picomoles (pmole)
1 mole	=	10e12 picomoles
A 1 micromolar solution (1uM)	=	1 picomoles per microliter (1pmoles/ul):
1uM	=	10e-6 moles/liter x 10[-6] liters/ul
	=	10e-12 moles/ul = 1 pmole/ul

2. Percent solutions

A 1% solution is defined as 1 gram per 100 ml volume. To find the weight (in grams) needed for a particular volume (in ml) of solution, convert the desired percentage to a decimal (divide by 100) and multiply by the ml final volume desired.

Example 1: To make 4 liters of 10% SDS:

$$\frac{10\%}{100} \times 4000 \text{ ml} = 400 \text{ g SDS}$$

In practice, dissolve 400 g SDS in about 3500 ml dH2O. After the powder is in solution, the volume of the solution is adjusted with dH2O to a final volume of 4 liters.

Example 2: To make 250 ml of 0.8% TA agarose:

$$\frac{0.8\%}{100} \times 250 \text{ ml} = 2 \text{ g agarose required.}$$

In practice (when the % solution is very low) we do not worry about the volume displacement and add 250 ml TA buffer to 2 g agarose.

3. Concentration calculations

$$\text{Volume of stock required} = \frac{\text{Concentration required} \times \text{total volume}}{\text{Stock concentration}}$$

Example 1: To make 10 ml of a 15 mM solution from a 500 mM stock:
15 mM x 10 ml volume of stock required = ________________ = 0.3 ml or 300 ul
500 mM
Example 2: To make 10 ml of a 0.1% solution from a 20% stock:
0.1% x 10 ml
Volume of stock required = ________________ = 0.05 ml or 50 ul
20%

4. X solutions (X as in Times, e.g., 20X SSC)

A 1X solution is a solution in which all the components are at the standard working concentrations. For example, restriction enzyme digests are carried out in a 1X restriction buffer (the working concentration). However, the restriction buffers are prepared at a 10-fold increase in concentration (10X buffers, see Restriction Buffers section for examples), and diluted to 1X for use in restriction digests. A concentrated solution is *usually* diluted to 1X for use. An exception to this is 20X SSC which is used at 20X, at 10X (diluted 1:1 with dH2O), at 2X SSC (dilution: 100 ml 20X SSC+ 900 ml dH2O), etc.....

5. Determining pmoles of ends available for end labeling

A 1uM solution = 1pmole/ul; therefore a 25 uM oligonucleotide solution has 25 pmoles oligo/ul. If 10 pmoles of ends are required for a reaction, the 25 uM solution is diluted to prepare a 10 uM solution by following steps in #3 above: 4 ul of the 25 uM solution is diluted to a final volume of 10 ul = 10 uM solution Each ul of this dilution contains 10 pmoles of fragment; because the molecules are oligonucleotides there would be one end of each molecule available for endlabeling with kinase. If the molecules were double stranded DNA fragments, this diluted solution would contain 20 pmoles of ends/ul available for kinasing.

6. How to calculate the molar concentration of template (& molecules of each) Definitions:

(1) MW of template = length of template in bp x 660 daltons/bp
(2) a 1 x 10^6 dalton molecule is 1 ug/ pmole
(3) 1 mole has 6.02 x 10^{23} molecules = 10[12] pmoles
(4) 1 pmole = 6.02 x 10^{11} molecules

MAKING DILUTIONS

Many of you appear to panic when you must dilute something, yet the mathematics involve nothing worse than the simplest algebra. One reason is simply that when you are busy with a laboratory procedure you are distracted and it is difficult to think in the abstract. That problem can be overcome by practicing in advance of the need. Even with practice, though, you may find dilution problems confusing unless you very clearly define your objectives. We will give you a useful formula for making dilutions, one that you may have seen before. The formula is worse than useless, though, if you don't use it properly.

ESTABLISH A FRAME OF REFERENCE

For the sake of simplicity, lets say we are talking about sucrose solutions. Suppose you have a starting solution of sucrose (in water) with volume V1 and concentration C1. What is the total amount of sucrose in your solution? Answer: C1•V1.

Example 4: Volume = 0.2 liter; concentration is 50 grams/liter. C1•V1 = 50 grams/liter • 0.2 liter= 10 grams sucrose.

Now suppose that you dilute that solution with water—the whole thing—to some larger, predetermined volume (V2). What amount of sucrose is present in the new, diluted solution? If you said 10 grams, you get the gold star. But wait a minute—C1•V1 = 10 grams, and the new solution has a different volume, V2. The same amount of sucrose is present in the new solution as was in the original solution, so the following relationship must hold:

$C1 \times V1 = C2 \times V2$, where C2 = concentration of the new solution.

Example 5: Dilute the previous sucrose solution to 2 liters. What is the concentration of the new solution? We must solve for C2, of course. $C1 \cdot V1 = C2 \cdot V2 = 10$ grams. We know that V2 = 2 liters, so now we have

$C2 \cdot (2 \text{ liters}) = 10$ grams

Solve for C2 to obtain 5 grams/liter.

Determining what you already knows and putting the informatin into the equation $C1 \cdot V1 = C2 \cdot V2$ establishes the relationship that you need in order to solve dilution problems.

DETERMINE THE OBJECTIVE

What do you want to do? Or, more realistically, what does the instructor want you to do? Two types of diution problems are quite common in biology and biochemistry labs.

- Dilute a known volume of known concentration to a desired final concentration
- Dilute a known concentration to a desired final concentration AND volume

The second type of problem really throws people off! Let's start with the first one, though. You know V1, C1, and C2 is predetermined. It remains, then, to solve for V2, namely the final volume to which to dilute the solution. This one is easy, since you keep the amount of solute the same and only have to change one factor.

Now for the second type problem. You know C1, namely the concentration of the starting solution. You have predetermined both V2 and C2, namely the final volume and concentration that you desire. There is one undetermined variable left, namely V1. V1 is the volume of original solution that you will dilute to the desired final volume and concentration.

$$C1 \cdot V1 = C2 \cdot V2$$

$$V1 = (C2 \cdot V2)/C1 \text{ or,}$$

$$V1 = (\text{final amount of solute})/C1$$

Example 6: You have a sucrose solution of 47 grams/liter. You want to prepare 100 milliliters (0.1 liter) of sucrose solution of concentration 25 grams/liter. Since you know the starting concentration of sucrose and you know both the final concentration and volume of solution that you want, all you need to find out is what volume of starting solution (V1) to use.

$$V1 = (C2 \cdot V2/C1), \text{ that is,}$$

$$V1 = (25 \text{ grams/liter} \times 0.1 \text{ liter}) \div 47 \text{ grams/liter} = 0.053 \text{ liter} = 53 \text{ ml}$$

Notice that the calculation comes out to 0.05319149L using full precision, but I rounded off the required volume to the nearest ml. There is a limit to the precision with which we can prepare a solution, and also a limit to the precision that we really need. You would use a 100 ml graduated cylinder to determine final volume. You would be able to read the markings to the nearest 1 ml, as a rule.

UNITS

It is critical that you report units for concentrations, volumes, and amounts, and when you make calculations for dilutions you must not mix up the units. For example, it doesn't work to write, V1 (160 milliters)•C1(160 milligrams/liter = V2 (unknown)•C2(desired-3 grams/liter). However, because concentration represents a proportional relationship, you can select from a variety of units. For example, 1 milligram/milliliter is the same as 1 gram/liter, 1 microgram/microliter, or 1 nanogram/nanoliter. It is also the same as 1000 milligrams/liter, but why would we write it that way?

Select units that simplify your expressions.

WORKING WITH STOCK SOLUTIONS

We define a *stock solution* as a concentrate, that is, a solution to bc diluted to some lower concentration for actual use. We may use just the stock solution or use it as a component in a more complex solution. We refer to the solution that we end up using as a *working solution*. If you are comfortable making dilutions then you can appreciate the many advantages of working with stock solutions. Although it is never absolutely necessary to use a stock solution, it is often impractical *not* to use them. Stock solutions can save a lot of time, conserve materials, reduce needed storage space, and improve the accuracy with which we prepare solutions and reagents. Here are several illustrated types of applications using stock solutions.

WHAT DO THE RATIOS MEAN?

Suppose someone asks you to "prepare a one to ten dilution of solution X." Does it mean take one part solution X and add ten parts water, or does it mean take one part solution X and bring the volume to a total of ten parts? A biologist would likely apply the second definition to a buffer or reagent solution. In another discipline, though, the former definition might be more relevant. Even in biology, we often prepare complex media as weight-to-volume (w:v) instead of weight-in-volume (w/v), that is, we add a prescribed mass of material to a prescribed volume of water instead of mixing the materials and bringing the mixture to a prescribed final volume.

To avoid confusion you might say "please prepare a one to ten dilution of solution X, weight-in-volume," or if you want to bring materials together in a precise proportion, say "please prepare a one to ten dilution of solution X, weight-to-volume." In the latter case it might be less confusing just to say "please combine one part solution X with ten parts water."

Calculations

The concentration units for chemical standard solutions used for ICP applications are typically expressed in μg/mL (micrograms per milliliter) or ng/mL (nanograms per milliliter). For example, a 1000 μg/mL solution of Ca^{+2} contains 1000 micrograms of Ca^{+2} per each mL of solution and a 1 μg/mL solution of Ca^{+2} contains 1000 ng of Ca^{+2} per milliliter of solution. To convert between metric concentration units the following conversions apply:

Mass portion of concentration unit where g = gram

Prefix	Scientific Notation	Decimal equivalents	Example Units
kilo–(k)	$= 10^3$	$= 1000$ g	kilogram (kg)
milli–(m)	$= 10^{-3}$	$= 0.001$ g	milligram (mg)
micro–(μ)	$= 10^{-6}$	$= 0.000001$ g	microgram (μg)
nano–(n)	$= 10^{-9}$	$= 0.000000001$ g	nanogram (ng)
pico–(p)	$= 10^{-12}$	$= 0.000000000001$ g	picogram (pg)

Volume portion of concentration unit where L = liter

Prefix	Scientific Notation	Decimal equivalents	Example Units
milli–(m)	$= 10^{-3}$	= 0.001 L	milliliter (mL)
micro–(μ)	$= 10^{-6}$	= 0.000001 L	microliter (μL)
nano–(n)	$= 10^{-9}$	= 0.000000001 L	nanoliter (nL)
pico–(p)	$= 10^{-12}$	= 0.000000000001 L	picoliter (pL)

The difference between ppm and μg/mL is often confused. A common mistake is to refer to the concentration units in ppm as a short cut (parts per million) when we really mean μg/mL. One ppm is in reality equal to 1 μg/g. In similar fashion ppb (parts per billion) is often equated with ng/mL. One ppb is in reality equal to 1 ng/g.

Parts Per Million (PPM)

A unit of concentration often used when measuring levels of pollutants in air, water, body fluids, etc. One ppm is 1 part in 1,000,000. The common unit **mg/liter** is equal to ppm. Four drops of ink in a 55-gallon barrel of water would produce an "ink concentration" of 1 ppm.

Parts Per Billion (PPB)

One part per billion is 1 part in 1,000,000,000. One drop of ink in one of the largest tanker trucks used to haul gasoline would represent 1 ppb.

The difference between 1 ppm and 1 ppb is important. A prestigious scientific journal recently reported the concentration of a substance as 0.5-1.5 ppm. The real value was 0.5-1.5 ppb. The difference between $1 and $1000!

Parts Per Trillion (PPT)

A unit of concentration used to measure vanishingly small levels of pollutants or contaminants in, for example, body fluids. One ppt is 1 part in 1,000,000,000,000. One drop of ink distributed through the water contained in a total of 4 of the 3-million-gallon reservoirs pictured would result in a final concentration of 1 ppt.

The remarkable advances in the sensitivity of modern analytical techniques makes it possible to detect some substances at the ppt level whose presence would not have been detected using earlier assay methods.

To convert between ppm or ppb to μg/mL or ng/mL the density of the solution must be known. The equation for conversion between wt./wt. and wt./vol. units is:

$$(mg/g) \text{ (density in g/mL)} = mg/mL$$

Therefore, if we have a solution that is 1000 mg/mL Ca^{+2} and know or measure the density to be 1.033 g/mL then the ppm Ca^{+2} = (1000 mg/mL) / (1.033 g/mL) = 968 mg/g = 968 ppm.

When making dilutions the following equation is useful:

$$(mL_A)(C_A) = (mL_B)(C_B)$$

For example, to determine how much of a 1000 μg/mL solution of Ca^{+2} required to prepare 250 mL of a 0.3 μg/mL solution of Ca^{+2} we would use the above equation as follows:

$$(mL_A)(1000 \text{ mg/mL}) = (250 \text{ mL})(0.3 \text{ mg/mL})$$

$$(mL_A) = [(250 \text{ mL})(0.3 \text{ µg/mL})]/ (1000 \text{ µg/mL})$$

$$(mL_A) = 0.075 \text{ mL} = 75 \text{ µL}$$

PREPARATION OF STANDARD CHEMICAL SOLUTIONS

Standard chemical solutions can be prepared to weight or volume. The elimination of glass volumetric flasks may be necessary to eliminate certain contamination issues with the use of borosilicate glass or to avoid chemical attack of the glass. It is often assumed that 100 grams of an aqueous solution is close enough to 100 mL to not make a significant difference since the density of water at room temperature is very close to 1.00 (0.998203 at 20.0 °C). Diluting / preparing standard solutions by weight is much easier. Still, the above assumption should not be made. The problem is that trace metals standards are most commonly prepared in water + acid mixtures where the density of the common mineral acids is significantly greaten than 1.00. For example, a 5% v/v aqueous solution of nitric acid will have a density of ~1.017 g/mL which translates into a fixed error of ~1.7%. Higher nitric acid levels will result in larger fixed errors. This same type of problem is true for solutions of other acids to a degree that is a function of the density and concentration of the acid in the standard solution as described by the following equation (to be used for estimation only):

$$d_S = [(100–\%) + (d_A)(\%)] / 100$$

Where:

d_S = density of final solution

% = The v/v % of a given aqueous acid solution

d_A = density of the concentrated acid used

For example, lets estimate the density of a 10% v/v aqueous solution of nitric acid made using 70% concentrated nitric acid with a density of 1.42 g/mL.

$D_S = [(100-\%) + (d_A)(\%)]/100 = [(100-10) + (1.42)(10)]/100 = (90 + 14.2)/100 = 1.042$ g/mL

ACID CONTENT

Another area of confusion is the expression of the acid content of the solution. We all agree that it is important to matrix match the standard and sample solutions to avoid a fixed error in the solution uptake rate and/or nebulization efficiency sometimes referred to as a matrix interference. If a solution is labeled as 5% HNO_3 what does this mean? If we take 5 mL of 70% concentrated nitric acid and dilute to a volume of 100 mL then this is 5% HNO_3 (v/v) where the use of 70% concentrated acid is assumed. However, nitric acid can be purchased as 40%, 65%, 70%, and > 90%. Therefore, note the concentration of the concentrated acid used if different from the 'norm' as well as the method of preparation i.e. v/v or wt/wt or wt/v or v/wt. The wt. % concentrations of the common mineral acids, densities, and other information are shown in the following table:

Weight % Concentrations

Acid	Mol.Wt.	Density (g/mL)	Wt. %	Molarity
Hydrochloric	36.46	1.19	37.2	12.1
Hydrofluoric	20.0	1.18	49.0	28.9
Nitric	63.01	1.42	70.4	15.9
Perchloric	100.47	1.67	70.5	11.7
Phosphoric	97.10	1.70	85.5	14.8
Sulphuric	98.08	1.84	96.0	18.0

ACID CONTENT IN MOLARITY

It is important to know what the concentration units of the concentrated acid being used mean. Taking 70% concentrated nitric acid as an example means that 100 grams of this acid contains 70 grams of HNO_3. The concentration is expressed at 70% wt./wt. or 70 wt. % HNO_3. Some analysts prefer to work in matrix acid concentrations units of Molarity (moles/liter). To calculate the Molarity of 70 wt. % nitric acid we calculate how many moles of HNO_3 are present in 1 liter of acid. Lets say that we tare a 1 liter volumetric flask and then dilute to the mark with 70.4 wt. % HNO_3. We would then measure the weight of the solution to be 1420 grams. Knowing that the solution is 70.4 wt % would then allow us to calculate the number of grams of HNO_3 which would be (0.704)(1420g) = 999.7 grams HNO_3 per liter. Dividing the grams HNO_3 by the molecular weight of HNO_3 (63.01 g/mole) gives the moles HNO3 / L or Molarity which is 15.9 M. The above logic explains the following equation used for calculating the Molarity of acids where the concentration of the acid is given in wt %:

$$[(\% \times d) / MW] \times 10 = \text{Molarity}$$

Where:

$$\% = \text{wt. \% of the acid}$$

$$d = \text{density of acid (specific gravity can be used if density not available)}$$

$$MW = \text{molecular weight of acid}$$

Using the above equation to calculate the Molarity of the 70 wt % nitric acid we have:

$$[(70.4 \times 1.42) / 63.01] \times 10 = 15.9 \text{ M}$$

Dilutions of the concentrated acid to prepare specific volumes of specified Molarity can be make using the $(mL_A)(C_A) = (mL_B)(C_B)$ equation.

MOLARITIES AND DILUTIONS

Technically, one mole is:

"The amount of a substance which contains as many elementary entities as there are atoms in 0.012 kilogram of carbon 12".

It was long after Avogadro that the idea of a mole was introduced. Since a molecular weight in grams (mole) of any substance contains the same number of molecules, then according to Avogadro's principle, the molar volumes of all gases should be the same. The number of molecules in one mole is now called

Avogadro's number. Avogadro had no knowledge of moles, or of the number that was to bear his name, which was never actually determined by Avogadro himself.

Avogadro's number is: 602,000,000,000,000,000,000,000   (6.02×10^{23})

How big is Avogadro's number?

- An Avogadro's number of soft drink cans would cover the surface of the earth to a depth of over 200 miles.
- If you spread Avogadro's number of unpopped popcorn kernels across the USA, the entire country would be covered in popcorn to a depth of over 9 miles.
- If we were able to count atoms at the rate of 10 million per second, it would take about 2 billion years to count the atoms in one mole.

A **solution** is a homogeneous mixture where all particles exist as individual molecules or ions. The **molarity** of a solution is calculated by taking the moles of solute and dividing by the litres of solution:

$$\text{Molarity} = \frac{\text{moles of solute}}{\text{litres of solution}}$$

Just to confuse you, the **molality** (m) (yes, that is spelt right) of a solution is the concentration measured as moles of solute per kilogram of solvent. For example, a 1 m (not a 1 M) NaCl solution contains 1 mole of NaCl per kilogram of water. Molalities are preferred over molarities in experiments that involve temperature changes of solutions, e. g. calorimetry and freezing point depression experiments. Molarity = moles per litre, molality = moles per kilogram. OK, we won't talk about molality any more.)

Example 1: What is the molarity of 2 moles of solute dissolved in 1 litre of solvent?

$$\text{Molarity} = \frac{2 \text{ mol}}{1 \text{ litre}} = 2 \text{ mol} / \text{L} = 2\text{M}$$

i.e. 2 moles/litre, or "2 molar".

Example 2: What is the molarity of 0.75 moles of solute dissolved in 2.5 litres of solvent?

$$\text{Molarity} = \frac{0.75 \text{ mol}}{2.5\text{L}} = 0.3\text{M}$$

Example 3: What is the molarity of 40 grams of NaOH dissolved in 2 litres of solvent? This calculation must be performed in two stages:

   a. Convert grams to moles
   b. Divide moles by litres to calculate molarity

a) The molecular weight of NaOH is 40 grams/mol

$$\frac{40\text{g}}{40\text{g/mol}} = 1\text{mol}$$

$$\frac{1\text{mol}}{2\text{L}} = 0.5\text{M}$$

Equations to be remembered

1. $$\frac{\text{Grams}}{\text{Molecular weight}} = \text{moles (mol)}$$

2. $\dfrac{\text{Moles}}{\text{Volume}}$ =   molarity (M)

3.   Molarity (M)   = $\dfrac{\text{Grams}}{\text{Molecular weight}}$

Example 4: When 2 grams of NaCl (molecular weight 58.44 g mol^{-1}) is dissolved in 100 mL of solute, what is the molarity of the solution? (MV=g/mol.wt.)

$$(x\ M)\ (0.1L) = \dfrac{2g}{58.44}$$

Therefore:

$$(' M)\ (0.1) = 0.034 = 0.34M$$

Example 5: How many grams of NaCl are needed to make 500 ml of a 0.2 M solution? (MV=g/mol.wt.)

$$(0.2\ M)\ (0.5L) = \dfrac{x}{58.44}$$

Therefore:

$$0.1 = \dfrac{x}{58.44} = 5.844$$

DILUTION CALCULATIONS

Solutions are prepared sometimes by diluting a more concentrated solution; hence one has to be familiar with the calculations that are associated with dilutions. For example, if you needed a one molar solution you could start with a six molar solution and dilute it.

It is quite simple to calculate such types of equations and the element of simplicity is that the number of moles of solute stays the same, as shown here. The number of moles of solute in the concentrated solution (indicated by the subscripted moles$_{con}$) is equal to the number of moles in the dilute solution. You have simply increased the amount of solvent in the solution.

$$\text{moles}_{con} = \text{moles}_{dil}$$

Of course you know that the number of moles of solute in the concentrated solution is equal to the molarity of the concentrated solution times the volume of the concentrated solution. Also, the number of moles of solute in the dilute solution is equal to the molarity of the dilute solution times the volume of the dilute solution. Since we are really interested in the molarities and volumes we can substitute and use the equation shown in the second line (or third line, depending on your preference for how to show multiplication). Let's use this equation in a few examples.

$$\text{moles}_{con} = \text{moles}_{dil}$$

$$\text{moles}_{con} \times \text{Vol}_{con} = M_{dil} \times \text{Vol}_{dil}$$

$$(Mcon)\ (V_{con}) = (M_{dil})\ (V_{dil})$$

DILUTION CALCULATIONS: PRACTICE

1. How much 2.0 M NaCl solution would you need to make 250 mL of 0.15 M NaCl solution?
   Ans. 19 ml (2 s.d. from 18. 75ml)

2. How much 2.0 M NaCl solution would you need to make 250 mL of 0.15 M NaCl solution?

   Ans. 0.76 M (2 s.d.)

3. What would be the concentration of a solution made by adding 250 mL of water to 45.0 mL of 4.2 M KOH?

   Ans. 0.64 M

4. How much 0.20 M glucose solution can be made from 50 mL of 0.50 M glucose solution?

   Ans. 130 mL (2 s.d. from 125 mL)

Answers

Here are the answers to the questions in exercise

$$(M_{con}) (V_{con}) = (M_{dil}) (V_{dil})$$

$$(V_{con}) = \frac{(M_{dil})(V_{dil})}{(M_{con})}$$

$$(V_{con}) = \frac{(0.050\,M)(500.\,mL)}{(6.0M)}$$

$$V_{con} = 4.2\ mL$$

Chemist starts with 50.0 mL of a 0.40 M NaCl solution and dilutes it to 1000. mL. What is the concentration of NaCl in the new solution?

$$(M_{con}) (V_{con}) = (M_{dil}) (V_{dil})$$

$$(0.40\ M) (50.0\ mL) = (M_{dil}) (1000.\ mL)$$

$$(0.40\ M) (50.0\ mL) = (M_{dil})$$

$$(1000.\ mL)$$

$$0.020\ M = M_{dil}$$

Summary

- The concentration of a solution is usually given in moles per litre (mol L^{-1} OR mol/L).
  This is also known as *molarity.*
- Concentration, or Molarity, is given the symbol M.
  A short way to write that the concentration of a solution of hydrochloric acid is 0.01 mol/L is to write [HCl]=0.01M
  The square brackets around the substance indicate concentration.

- $M = n \div V$ (M=concentration of solution in mol/L, n=moles of substance, V=volume of solution in litres (L))
  This formula can be re-arranged:
- $n = M \times V$ (M=concentration of solution in mol/L, n=moles of substance, V=volume of solution in litres (L))
- $V = n \div M$ (M=concentration of solution in mol/L, n=moles of substance, V=volume of solution in litres (L))

Examples

1. $M = n \div V$

Calculate the concentration (molarity) of a sodium chloride solution containing 0.125 moles sodium chloride in 0.5 litres of water.

- $M = n \div V$ in litres
- $n = 0.125$mol
- V in litres = 0.5L
- $[NaCl(aq)] = M = 0.125 \div 0.5 = 0.25M$ (or 0.25mol/L or 0.25mol L^{-1})

2. $n = M \times V$

Calculate the moles of copper sulfate in 250mL of 0.02M copper sulfate solution.

- $n = M \times V$
- $M = 0.02M$
- $V = 250mL = 250 \div 1000 = 250 \times 10^{-3}L = 0.250L$ (since there are 1000mL in 1L)
- $n = 0.02 \times 250 \times x\ 10^{-3} = 0.005$mol

3. $V = n \div M$

Calculate the volume of a 0.80M potassium bromide solution containing 1.6 moles of potassium bromide.

- $V = n \div M$
- $n = 1.6$mol
- $M = 0.80M$
- $V = 1.6 \div 0.80 = 2.00L$

IMMUNOLOGY

4

PREPARATION OF ANTIGEN

AIM

To prepare Antigen

- **Soluble: - BSA (Bovine Serum Albumin)**
- **Insoluble: -sheep RBC**
- **Particulate (whole organism): -Pathogen**

PRINCIPLE

The molecule that reacts or binds with antibody is called an antigen. While that antigen which not only binds but also induces an immune response is called an immunogen.

Characteristics of an Immunogen

1. *Molecular weight:* Generally high molecular weight can act as immunogen, e.g. BSA (66kD), Ovalbumin (42kD). Small molecular weight molecules when coupled with a carrier, BSA acts as an immunogen. Such small molecular weight molecules are called Haptens.
2. The immunogen should be a foreign substance so that it is not recognized as 'self' by the host organism.
3. *Chemical Structure:* To induce a strong immune response against a particular antigen, it is required that antigen has the capability to induce more than one type of antibodies. If for example a polysaccharide with homopolymer structure is injected, it will only trigger one type of antibody. Thus a weak response will be seen.
4. *Degradability:* The immunogen should be stable for sufficient time enough to induce a strong antibody reaction. If the foreign molecule is degraded very fast then enough antibodies will not develop so as to get detected easily.

CLASSIFICATION OF ANTIGENS

Based on the source from which antigen originally originates, they are classified in the following manner:

1.	Natural Antigen	:	RBC
2.	Exogenous Antigen	:	Pollen, Microbes
3.	Endogenous Antigen	:	Within the host, e.g. Mitochondria
4.	Artificial Antigen	:	Hapten.

Among these categories various types of antigens can be listed:

5.	Soluble Antigen	:	BSA
6.	Insoluble Antigen	:	Pollen, Cells (RBC)
7.	Whole organism Antigen	:	Bacteria

A. Preparation of Soluble Antigen: BSA

Quantity to be prepared	=	50ml
Amount of BSA per ml	=	0.5mg

To be prepared in Phosphate Buffered Saline (PBS)

B. Preparation of PBS

NaCl	=	8gm
KCl	=	0.2gm
KH_2PO_4	=	0.2gm
$Na_2HPO_4 . 2H_2O$	=	1.15gm
PH	=	7.2

Prepare 50ml. To this add 25mg of BSA

C. Preparation of Insoluble Antigen: Sheep/Goat RBC

Anticoagulant solution (Alsever's solution)
Quantity to be prepared   =   100ml

Components of Alsever's solution

Dextrose	=	20.5gm
Trisodium Citrate	=	8.0gm
NaCl	=	42.0gm
Citric Acid	=	0.55gm
Water	=	100ml

D. Preparation of RBC

1. Collect sheep or goat blood in a sterilized bottle
2. Add Alsever's solution (1ml to every 5ml of blood) and shake
3. Centrifuge at 2000rpm at room temperature fro 5 min.
4. Discard supernatant and resuspend pellet in PBS
5. Centrifuge at 2000 rpm for 10 min at room temperature
6. Suspend in PBS. For prolonged storage add 1% Sodium Azide

E. Preparation of Antigen as Whole Organism

1. Use any microbe at a concentration of 10^6 per ml
2. Heat inactivate by keeping at 65° C for 10 min. or treat with 5% formaldehyde

5

AIM OF THE EXPERIMENT

To immunize mice and rabbit using different kinds of antigen

PRINCIPLE

A. The Immune Response

An immune response is the consequence of a complex sequence of events involving antigen and at least three kinds of lymphoid or reticuloendothelial cells. B-cells, in response to antigens, produce antibodies in blood. A further complexity is introduced into the system by the different classes (isotopes) of antibody molecules produced by 'B' cells (fig) and the past immunization history of the animal which is reflected by the number of 'B' memory cells present. A high number of memory cells results in a very large and much more rapid response to a second injection (booster dose) of antigen -a secondary response- classical feature of the immune response.

B. Parameters affecting Immune Response

Antigen Dose

It is clear that the nature of the antigen introduced partially into an animal is important. However the amount is of greater significance. Too little or too much antigen can be striking counter productive from the point of view of the immunologist trying to obtain a maximum yield of antibody. Optimum dosage for various experimental animals is as follows:

Animal	Dosage
Mice	5-50µl
Rabbit	50-500µl
Rats	10-100µl
Guinea Pigs	10-100µl
Sheep/Goat	0.5-200µl
Chicken	50-200µl

AGE OF THE ANIMAL

Immune response is strikingly age dependent. Newborn animals having next to no innate ability to synthesize antibody and old mammals showing a response very significantly reduced by comparison with that seen in their prime age. In new born babies maternal antibody is transferred passively either across the placenta or orally through the colostrums or milk. Such passively acquired antibody can interfere with the process of active immunization

GENETIC FACTORS

Certain antigens (Ag) can elicit an immune response in inbred strain but not another. This ability to response has in many cases been shown to be under genetic control of loci closely linked to or part of the MHC. Genetic factors are undoubtedly major sources of variation in an immune response when out bred animals are immunized.

GENERAL HEALTH

The general health of the animal used to raise an antiserum can have a significant effect on antibody (Ab) levels. Poor nutrition, disease-ridden animals or dirty housing and other stress related situations could be reflected in poor response.

ROUTE OF IMMUNIZATION

There are five routes through which antigen can be injected:

1. Intramuscular
2. Intravenous
3. Intradermal
4. Intraperitoneal
5. Subcutaneous

Specific routes are taken depending on the animals to be immunized and the type of antigen used. For example, intravenous injections are not given if the antigen is a particulate matter because otherwise it may cause a cross-reaction. Intraperitoneal injection are given only to small animals like mice, rats etc.

ANTIGEN AND ADJUVANT

The most important factor to be considered in immunization experiments is the type of antigens used. It is very important that an antigen should be an immunogen also, i.e. it should spark off an immune response. This having taken care of, antigen is chosen from antigen classes of proteins, polysaccharides, synthetic polypeptides, chemically modified antigens, low molecular weight substances, etc. All these vary in efficiency and speed of instigating an immune response.

An essential required condition for immunization is that the immunogen should degrade slowly in the body of the experimental animal. This is to make that immune system of the animal gets ample time to synthesize Ab against the Ag. To make the Ag last longer in the body Adjuvants are used. Various types of adjuvants used these days are:

1. Freunds Complete adjuvants (FCA): its components are

    a) Mineral oil Bayol – f 3
    b) Emulsifying agent Mannide Mono Oleate
    c) Muramyl Dipeptide components of the membrane of heat killed Mycobacterium

2. Freund's Incomplete Adjuvant (FIA): It is FCA only but without the bacterial component.
3. Insoluble Aluminum salt + Calcium
4. Heat killed Mycobacterium tuberculosis
5. Bacterial Lipopolysaccharide
6. Saponin
7. Synthetic polynucleotide, Lipopoly TC / Poly AU
8. Heat killed *Bordetella pertussis*

Saponin is safer, used in veterinary medication

RAISING ANTIBODIES IN MICE

RAISING ANTIBODIES IN MICE

Mice:

Depending on the strain of mice used, the Ab production varies. For example:

BALB / C Mice	Ab Production	Intermediate
CS7BL / 6	"	Low
CBA	"	
	"	High

Use Swiss strain of mice

MOUSE HANDLING (FIG. 2)

1. Remove the mouse from the cage by gently grasping the tail (with preferred hand) at the base.
2. Place the animal on a wire-bar cage lid to permit grasping.
3. Approach the back of the neck from the rear with free hand. Firmly grasp the skin behind the ears with the thumb and the index finger.
4. Transfer the tail from preferred hand to beneath the little finger of the hand the skin of the neck.
5. Observe or inject the restrained mouse.

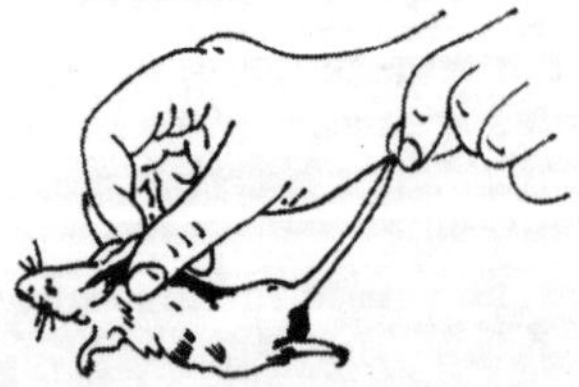

Fig. 2 Two Handed Method

Fig. 3 One Handed Method

ANIMAL IDENTIFICATION

Proper identification of animals plays an important role in research, whether animals are being used as experimental subjects or for breeding purposes, or as a source of tissue, cell or fluids.

Basic information is usually maintained on cage cards that should be utilized to identify single or group housed rodents where individual identification is not necessary. In some cases it is necessary to individually identify rodents.

Usually for mice ear puncturing is used for identification and for rabbit and goat ear tagging with numbered metal clips are used.

RAISING ANTIBODIES IN MICE

ANTIGEN: Sheep/Goat RBC (25%)

1. Inject 100µl of 25% s /g RBC to each animal either subcutaneous or intraperitoneally.
2. Label the animal by ear puncture and label also the cage.

PROTOCOL FOR INJECTION

1. Inject 100µl of 25% s /g RBC on day one through intraperitoneal route.
2. Give booster injection (100µl of 25% s /g RBC) on day 15 through subcutaneous route.

Bleed after 3 days by retro-orbital bleeding (fig) / heart puncture.

RAISING ANTIBODIES IN RABBIT

ANTIGEN: BSA

1. Take Freund's complete adjuvants in small flask (1.5ml).
2. To this add antigen BSA (also 1.5ml, 0.5mg. per ml) drop by drop and vortex simultaneously.
3. Mix adjuvant thoroughly before adding. This is done to uniformly distribute mycobacterium.
4. Inject 1ml into rabbit simultaneously.
5. Label the animal

PROTOCOL FOR INJECTION

1. First injection on 01 day, at the rate of 1mg/ml of BSA- subcutaneous injection with complete adjuvant.
2. Give the second injection on day 7, at the rate of 500µg/ml with incomplete adjuvant.
3. Booster injection can be given on day 28th. And 29[th.] at the rate of 500µg/ml with incomplete adjuvant.

Bleed after 5th day by ear bleeding/heart puncture.

NEGATIVE CONTROL

Maintain negative control by exposing another set of animals to some conditions, i.e. wounding with needle and injecting PBS without antigen.

Collect pre-immune serum for all animals before immunization by ear bleeding the rabbit and tail bleeding the mice

7

BLOOD COLLECTION AND SERUM PREPARATION

Serum is prepared by collecting blood samples from each immunized animals and also from the negative controls.

BLOOD COLLECTION FROM MICE

1. Manually restrain the animal.
2. Hold the animal in such a way that eyeball protrudes out.
3. Take sterile Pasteur pipette (or with a capillary tube with smooth edged end). Introduce it at the medial canthus of the orbit as shown in the figure.
4. Slowly, and with axial rotation, advance the tip of the pipette gently towards the rear of the socket until blood flows into the tube.
5. Collect the blood as it flows into the pipette by capillary action.
6. Blow the blood into a pre-labelled tube.
7. Dot excess blood from orbital site with a gauze sponge or swab moistened in saline or PBS.

BLOOD COLLECTION FROM RABBIT

1. Shave a small area near the edge of the ear.
2. Make a small cut in the ear vein and collect blood.
3. Use xylene (vasodilator) for free flow of blood if needed.
4. Apply antiseptic powder after bleeding.

SERUM PREPARATION

1. Allow the tubes to stand at room temperature for 30 min., disturb the clot and leave it at 4°C overnight.
2. Separate the clot by spinning at 2500 rpm for 10 min.
3. Transfer the serum to a separate tube.
4. Keep in water bath at 57°C for 30 min. to decomplement.
5. Add sodium azide to get a concentration of 0.1%.
6. Aliquot and store at – 20°C for further experiments.

8

AIM OF THE EXPERIMENT

To observe haemagglutination of RBC and to determine antibody titre for specific immunization.

PRINCIPLE

The basis of haemagglutination is the agglutination of RBC by the antibodies directed against them. Agglutination is the cross linking of the antigenic molecules by the antibodies to form an insoluble complex. Since RBC in the antigen agglutinated, it is called haemagglutination.

All RBC are multivalent, multideterminant particulate antigens. The RBC has a net negative charge on their surface due to presence of sialic acid. Thus a repulsive potential, which is a measure of the effective electrical electrostatic charge, is created called the beta potential. The lattice formation is dependent upon the overcoming of free receptor valance of antibodies to antigen between RBCs.

Haemagglutination Inhibition

A. POSITIVE REACTION

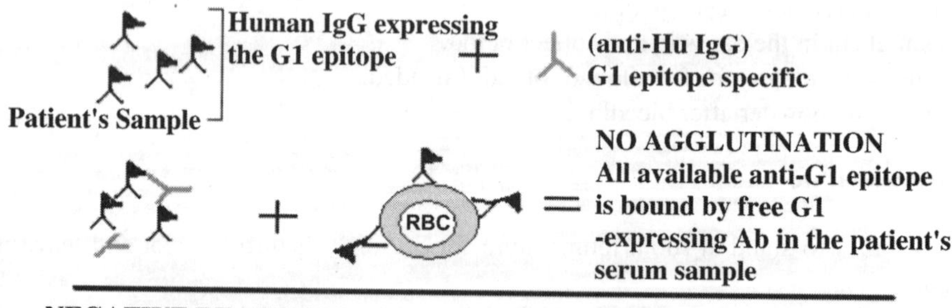

B. NEGATIVE REACTION

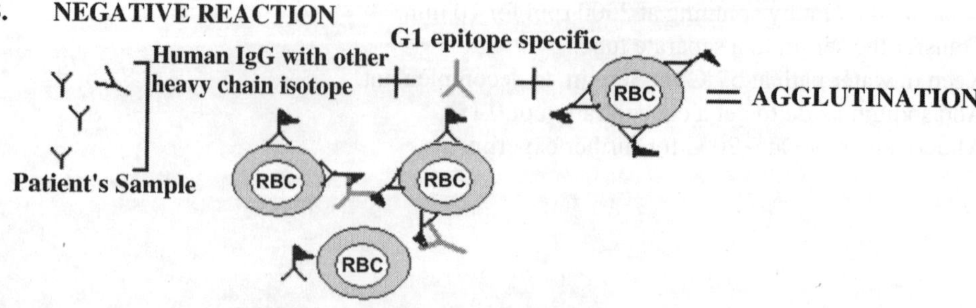

Fig. 4 Mechanism of haemagglutination

The antibodies capable of reacting with the antigen in saline solution are complete antibodies comprising mostly of IgM. IgM antibodies are pentameric and multivalent and are long enough to agglutinate RBCs. The antibodies incapable of reacting in saline are incomplete or blocking antibodies like the IgG molecules. IgG molecules are short and cannot agglutinate RBCs effectively though they get firmly attached to the antigens. The basic principle of haemagglutination is shown below:

The Ag – Ab concentrations should be in optimal ratio. High concentration of serum will show no agglutination due to the failure of lattice formation because of competition between the antibodies excess of prozone. If the antigen concentration is more also, agglutination not seen as little antibody combines with antigen, called zone of antigen only when the two concentrations are optimum- called zone of equivalence. So titre is the highest where there is visible agglutination.

TYPES OF HAEMAGGLUTINATION REACTIONS

a) **Direct:** Ab here is directed against the antigenic determinants or epitopes on the surface of RBC, which are the natural constituents of RBC.

b) **Indirect or Passive:** Ab is directed against substances which bind to the RBCs and agglutinate them. These artificial epitopes are substances like Polystyrene etc., or use of Ab directed against the primary Ab bound to the RBCs, ie. Anti RBCs.

COOMB'S TEST

This is a serological technique used to agglutinate RBCs using Ab of IgG isotype. This involves addition of an Ab directed against gamma globulin, which provides a bridge against the two Ab coated RBCs.

Procedure

1. Add 100 μl of PBS to all wells of Takatsy plate.
2. In the first well, add 50 μl of antibody against sheep(s) / goat (g) RBC and mix well.
3. Transfer 50 μl from the first well to the next in same row and repeat this situation in all the wells in first two rows. Discard the 50 μl from last well.

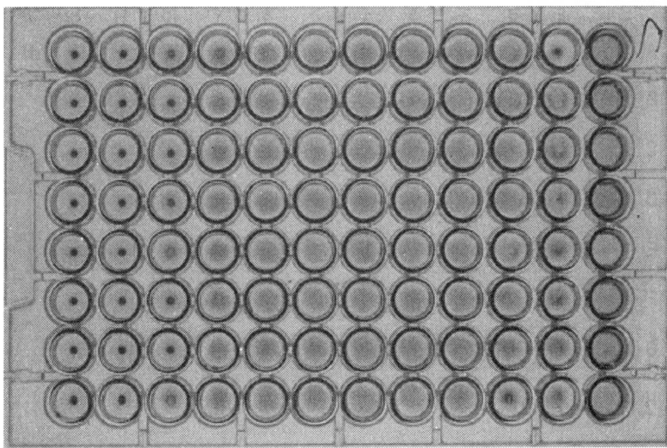

Fig. 5 Measurement of haemagglumtination by microtitre plate

4. Repeat the steps 1 – 3 with pre-immune serum and discard the last 50 µl.
5. Prepare s/g RBC suspension at a concentration of 1%.
6. Add 50 µl of 1% RBC to each well.
7. Shake the plate gently, cover the plate and incubate at 37°C for one hour and 4°C overnight.
8. Observe for agglutination, calculate the dilution and find out the titre (highest dilution which gave visible agglutination)
9. Compare the result with the control.

The dilution can be measured by colorimeter. The well containing the highest dilution which will give the visible agglutination of the sample.

Assay	Sensitivity*(µg antibody/ml)
1. Precipitin reaction in fluids	3 – 20
2. Precipitin reactions in gels:	
Mancini Radial Immunodiffusion	0.2 – 1.0
Double Immunodiffusion	3 – 20
Immunoelectrophoresis	3 – 20
Rocket Electrophoresis	0.2
3. Agglutination Reactions:	
Direct	0.05
Passive agglutination	0.001 – 0.01
Agglutination Inhibition	0.001 – 0.01
4. Radioimmunoassay	0.0001 – 0.001
5. Enzyme Linked Immunosorbent assay (ELISA)	0.0001 – 0.001
6. Immunofluorescence	1.0

*   The sensitivity depends upon the affinity of the antibody as well as the epitope density and distribution.

Source: '*Manual of Clinical Immunology*', N.R.Rose et al. (eds), 1986, American Society for Microbiology, Washington DC.

9

IMMUNO-ELECTROPHORESIS / IMMUNODIFFUSION

Antibodies are produced by the immune system in response to foreign macromolecules. Each antibody binds specifically to one feature (epitope) on one macromolecule (antigen). This allows the use of antibodies for the detection and quantitation of specific proteins in complex mixtures. Antibodies are generally isolated from animal serum, unless they are produced from tissue culture as monoclonals. The serum must be titrated to determine antibody concentration and specificity. Concentration determinations are often needed for antigen mixtures as well. Immunodiffusion and immunoelectrophoresis are useful techniques for these purposes.

If antibody and antigen are present in solution at approximately equal concentrations, they form an aggregate, which precipitates. This precipitate can be dissolved by the addition of an excess of antibody or of antigen (see figure below)

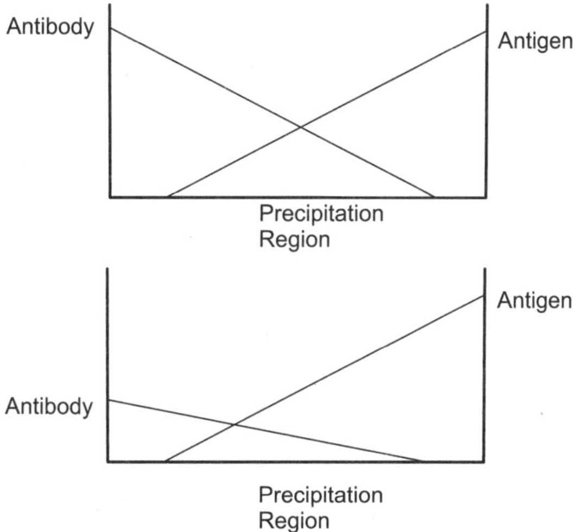

Fig. 6 Mechanism of precipitation when antibody reacts with antigen

When present at approximately equal concentration, antigen and antibody will precipitate as an aggregate. The figure above illustrates a principle underlying the usefulness of immunodiffusion that with shifts in antibody concentration, a corresponding shift in the region of precipitation will occur.

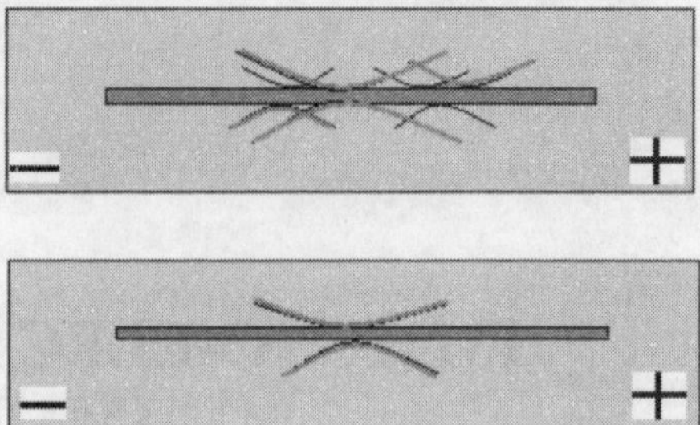

Fig. 7   Immunoelectrophoresis showing precipitin line

In a gel matrix, if a gradient is established in which antibody concentration decreases linearly in a given direction, while antigen concentration increases in the same direction, a precipitate will form a band perpendicular to this direction, at the point of approximately equal concentration. This is the basis of both immunodiffusion and immunoelectrophoresis. In both of these techniques, a gradient of antigen or of antibody and antigen, establishes a line of precipitate in the gel. The position of the line is indicative of the concentration of antigen or antibody in the sample, and may be calibrated against standards. The figures below show different variations of the immunodiffusion assay.

IMMUNOELECTROPHORESIS

AIM

To perform Immunoelectrophoresis (IE) using an antigen – antibody system.

PRINCIPLE

Immunoelectrophoresis combines separation of protein by electrophoresis with identification by double Immunodiffusion. An antigen mixture is first electrophoresed and separated by charge direction of the electric field and antiserum identifies individual antigen components. IE is widely used in clinical labs to detect the presence or absence of protein in the serum. The technique is applied for antigen, which migrate to the positive pole. It is used to detect Immunodeficiency diseases.

PROCEDURE

1. Prepare 1% agarose gel in barbital buffer (dissolve 4.4g **5'5 – diethylbarbituric** acid in 150ml of distilled water at 95°C. Make up to 900ml with cold water. Add 12g 5'5 – diethylbarbituric acid, sodium salt and adjust to pH 8.2 with sodium hydroxide. Make up to 1 litre) on a slide.
2. Punch wells and cut trough in the middle along the direction of the electric field. Do not remove the gel from the trough.
3. Load Ag into the wells and place in an electrophoretic chamber and electrophorase (pass current) after completing the connection by placing wicks (Whatman paper) at 80v.
4. After the run, remove agar from the trough and load antiserum.
5. Incubate the slide in a humid chamber overnight.
6. Stain the precipitin lines formed by placing in staining solution containing **Coommasie Blue G 250**. After 2 minutes destain the slide.

10

DOUBLE IMMUNODIFFUSION

AIM

To perform Ouchterlony's Double Immunodiffusion (DID) Assay.

PRINCIPLE

In this method both antigen (Ag) and antibodies (Ab) diffuse towards each other radially from the wells, thereby establishing a concentration gradient. As a result equivalence is reached, which is a visible line of precipitate formed.

This simple technique is an effective qualitative tool for determining the relationship between the Ag and the number of different Ab – Ag systems present.

PROCEDURE

In this method both antigen and antibody diffuse towards each other radially from the wells, thereby establishing a concentration gradient. The moment equivalence is reached, a visible line of precipitates forms. This simple technique is an effective qualitative tool for determining the relationship between the antigen and the number of different antibody – antigen systems present. The pattern of the precipitation line that forms when two different Antigen preparations are placed in adjacent wells indicates whether or not they share epitopes. For example, when two Ag share identical epitope, the antiserum will form a single precipitation line with both antigen that will grow towards each other and fuse to form a pattern called *'identity'*. If two antigens are unrelated, the antiserum will form independent precipitation line that cross, a pattern that establishes nonidentity. The lines cross because the unrelated Ag and Ab do not precipitate and are free to diffuse past other precipitation line forming across. If two Ag share the same epitope but one or the other has a unique epitope, a pattern of partial identity is formed.

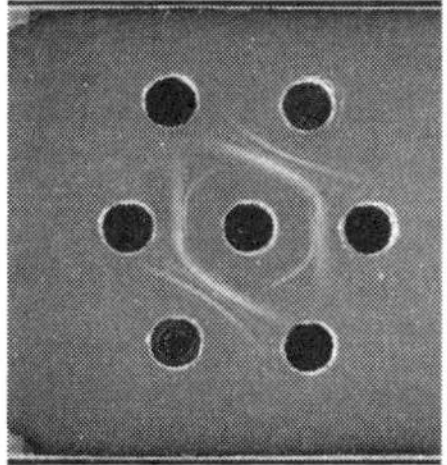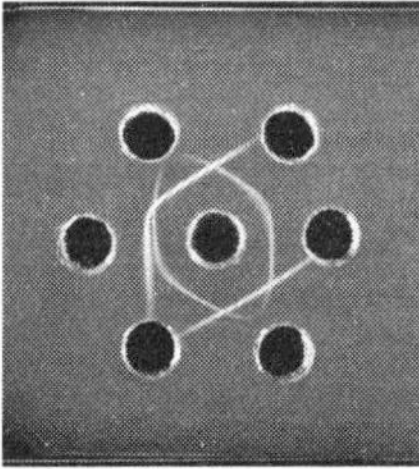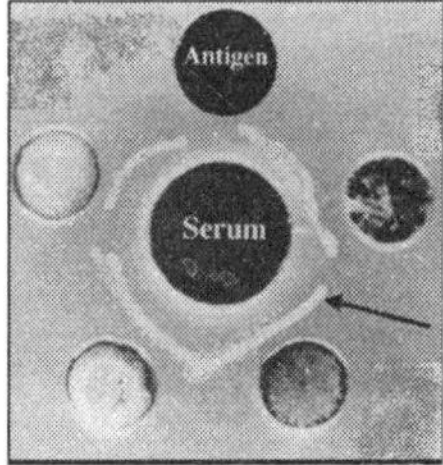

Fig. 8   The precipitin lines shown between wells containing Antigen and Antibodies

DID method, where antigens and antibodies diffuse toward each other, is very useful for establishing antigenic relationships between various substances. Three reaction patterns are seen in gel diffusion as follows.

Antibodies are detected by the precipitin reaction of DID method in agarose gel using animal organ extracts as antigen.

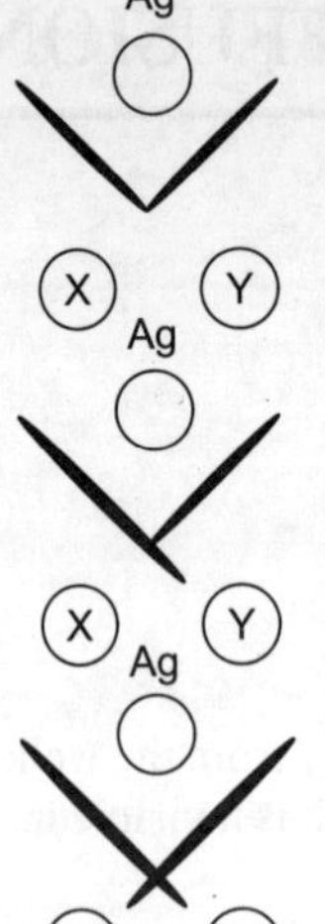

Fuse: When antibodies in X and Y recognize the same antigen, the two precipitin lines will fuse completely

Spur: When antibodies in X and Y partially recognize the same antigen, the two precipitin lines will form a spur

Cross: When antibodies in X and Y recognize two different antigens, the two precipitin lines will cross

Fig. 9   Types of precipitin lines

PROCEDURE

1. Prepare 10ml of 1% agarose by dissolving 0.10g in 10ml of PBS. Heat the solution to about 90°C to ensure complete dissolution of agarose.
2. Pour 3ml of the above solution on an alcohol cleaned glass slide using a pipette and allow solidifying.
3. Punch wells using a cork borer. Keep the slide in the fridge for 3 min. to ensure easy removal of agar from the wells.
4. Add 8 µl of Ab to the central well after the removal of agar wells and 8 µl of Ag in the peripheral wells.
5. If the Ab preparation has SDS, the slides should be kept outside in a humid chamber for 7hrs.
6. Recharge and allow remaining in the same atmosphere for 12 – 13 additional hrs.
7. Look for the formation of precipitation line occurred after 20 hrs and photograph.

11

SINGLE RADIAL IMMUNODIFFUSION

AIM

To perform Mancini's Single Radial Immunodiffusion (SRID)

PRINCIPLE

The Radial Immunodiffusion or the Mancini method is frequently used to determine the relative concentration of an antigen. It is a simple quantitative and qualitative assay in which an antigen sample is placed in a well and allowed to diffuse into agar containing a suitable dilution of antiserum. As the Ag diffuses into the agar, the region of equivalence is established and a ring of precipitation is formed around the well. The area of precipitin ring is proportional to the concentration of the Ag. By comparing the area of precipitin ring with a standard curve, the concentration of the Ag sample can be determined.

The Mancini technique is routinely used to quantitate serum levels of IgM, IgG, and IgA by incorporating class specific anti-isotope antibody into the agar. The technique is also applied to determine concentrations of complement components of serum. This method cannot detect antigens present in concentrations below 5 - 10μg /ml. This moderate sensitivity is the major limitation of the SRID.

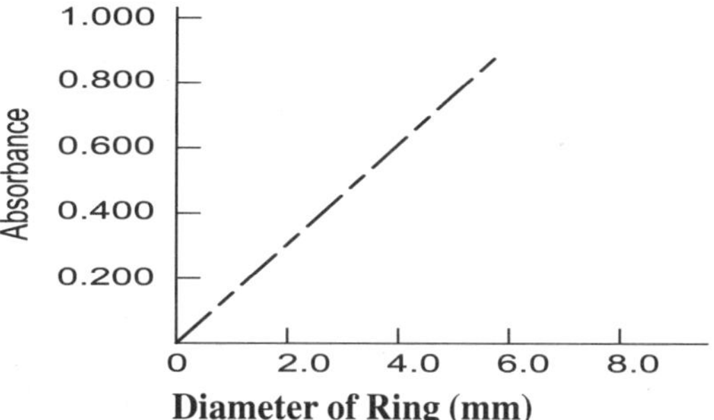

Fig. 10　Graph to calculate the diameter of the ring of precipitatin

PROCEDURE

1. Prepare 10ml of agarose by melting 0.1g of agarose in 10ml of PBS. To this add 100 μl of antiserum at about 45°C. Mix this thoroughly and pour onto slides.
2. After solidification punch wells in agarose and fill different wells with antigen.
3. Incubate the slide at room temperature in a humid chamber. Photograph the rings formed.

12

IMMUNOPRECIPITATION TECHNIQUE

AIM

To study the antigen – antibody interaction and assay by Immunoprecipitation technique.

PRINCIPLE

The principle of Immunoprecipitation is same as that of Immunodiffusion that is carried out in a solution. Precipitation is formation of an insoluble complex of antibody with different soluble antigen. If constant amount of Ab is added to various concentrations of Ag, a precipitate is formed in those mixtures, which corresponds to the zone of equivalence.

Quantitative analysis of this interaction gives both the Ab content of the immune serum and also an indication of the valency of the Ag, i.e. the number of dominant epitope clusters. Between the extremes of antigen and antibody excess, the cross linking of antigen and antibody will generally give rise to three dimensional lattice structures, which binds largely through Fc-Fc interaction, to form large precipitation aggregates.

PROCEDURE

1. Take 20 μl of serum in a series of Eppendorf tubes.
2. To these add increasing concentration of Ag (10 μl, 200 μl, 30 μl...............).
3. In all the tubes make volume up to 120 μl by adding PBS.

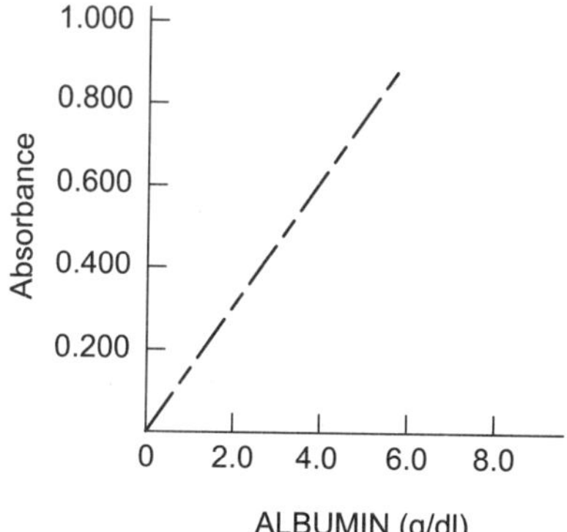

Fig. 11 Graph between absorbance and concentration to calculate the concentration of albumin in test samples

4. Incubate at 37°C for 1hr.
5. Transfer to 4°C over night.
6. Spin down at 10,000rpm for 10min. at 4°C.
7. Wash the pellets with cold PBS. Suspend the pellet in 100 µl. Centrifuge again discards the supernatant.
8. To the pellet add 10 µl. Of 0.1N NaOH. Make volume up to 1ml with distilled water.
9. Take O.D. at 260nm and 280nm or by an ELISA reader.
10. Take O.D. of first and last supernatant also to calculate the antibody precipitated.
11. Plot the graph.

13

ROCKET IMMUNOELECTROPHORESIS

AIM

To perform rocket Immunoelectrophoresis (**RIE**)

PRINCIPLE

In this technique, negatively charged Ag is electrophoresed in a gel containing Ab. The precipitate formed between Ag and Ab has the shape of a rocket, the height of which is proportional to the concentration of Ag in the well. One limitation of this technique is the need for negatively charged Ag. Some proteins, such as immunoglobulin do not have sufficient charge to be quantitated by RIE, nor is it possible to quantitate several Ag in a mixture at the same time.

Two- Dimensional Immunoelectrophoresis

Several Ag in a complex mixture can be quantitated simultaneously with a modification of RIE called 'Two Dimensional Immunoelectrophoresis' (2-RIE). In this technique Ag is first separated into components by electrophoresis. The gel is then laid over another agar gel containing antiserum, and electrophoresis is repeated at right angle to the first direction, forming precipitin peaks similar to those obtained with RIE (fig.10). Measurement of the size of the peaks allows quantitation of a number of proteins in a complex mixture.

PROCEDURE

1. Add antiserum to agarose gel, mix well and pour to a slide.

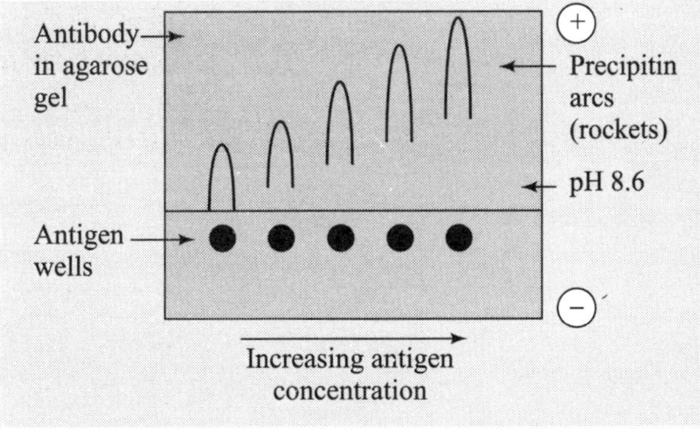

Fig. 12 Rocket-like shape of precipitate formed between Ag and Ab

2. Punch a well at one end of the slide and add Ag to it.
3. Electrophorase at 60v for one and a half hours.
4. Observe a rocket shaped precipitin line, stain, de-stain and photograph.

14

ENZYME LINKED IMMUNO-SORBENT ASSAY (ELISA)

AIM

To perform Enzyme Linked Immunosorbent Assay

PRINCIPLE

It is an extremely scientific technique, which does not require any precautions as against radioimmunoassay. It is more stable than labelled (radioactive) iodinated proteins. It is easy to handle, simple and multiple assays can easily be done.

The assay for an antigen depends on being able to couple highly specific antibody to enzyme and a solid substrate such as plastic beads or plates (the inner walls where assay is done)

The Ag is immobilized on a solid surface. The Ab specific to this Ag is added to this plate, which binds to the Ag. Then an Ab conjugation is done with the Ag again. This binds to Ab portion of the first or primary Ab. This complex can be detected by enzyme substrate reaction. The intensity of this chromogenic reaction is measured.

Generally, the polystyrene plates are employed whose binding capacity is 300ng/cm. The binding of the protein with the polystyrene is not constant – may be electrostatic as polystyrene has positive charges.

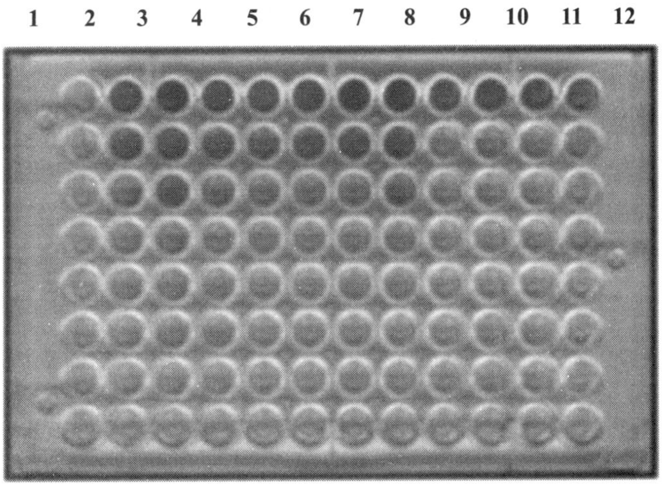

Fig. 13   An ELISA Plate with wells showing Ag–Ab reaction at various dilutions

Following are various modification of the basic ELISA technique

1. Direct ELISA

In this case the first Ab itself is conjugated with an enzyme, which binds specifically to the Ag. This complex is then assayed with the help of a chromogenic substrate reaction.

2. Indirect ELISA

The Ab conjugated with enzyme is directed against another Ab, which specifically binds to the Ag in use. This method is frequently used as it avoids unnecessary wastage of the primary Ab – enzyme conjugate.

3. Sandwich ELISA

The plate is first coated with the Ab, which binds through its Fc receptors. The antigenic receptors bind to the Ag added. The second Ab, conjugated with enzyme, is added after the addition of another Ab against Ag, but of different origin, which also specifically binds to the same Ag. The complex is then assayed using chromogenic reaction.

4. Competitive ELISA

In this technique Ab is first incubated in solution with the sample containing Ag. The Ag and Ab mixture is then added to an antigen coated microtitre plate. Addition of an enzyme conjugated secondary Ab can be used to quantitate the amount of primary Ab.

Enzymes and Substrates Available for ELISA

A number of enzymes have been employed for ELISA including Alkaline Phosphatase, Horse Radish Peroxidase and p-Nitro phenyl Phosphatase.

Enzyme		Substrate
Phosphatase	:	p-Nitro phenyl Phosphate 1mg/ml in Carbonate Buffer containing 5mM Magnesium Chloride
Peroxidase	:	Ortho Phenyl Diamine (OPD) 5mg in 12.5ml of Carbonate Buffer, add 5 µl of 30% Hydrogen peroxide.

ELISA PROTOCOL

There are many different types of ELISAs, which can detect the presence of protein in serum or supernatant. One of the most common types of ELISA is the so-called "sandwich ELISA." It is termed this

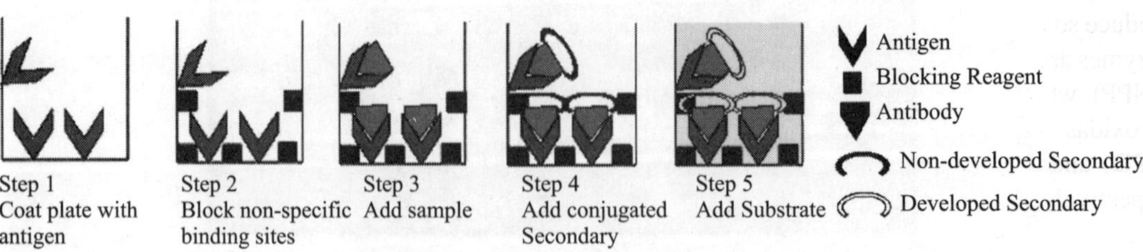

Fig. 14   ELISA Protocol

because the antibody that you are detecting gets sandwiched between an antigen and a chromogenically conjugated antibody. Below you will find a basic protocol for this assay.

Enzyme-linked Immunosorbent Assays (ELISAs) combine the specificity of antibodies with the sensitivity of simple enzyme assays, by using antibodies or antigens coupled to an easily assayed enzyme that possesses a high turnover number. ELISAs can provide a useful measurement of antigen or antibody concentration.

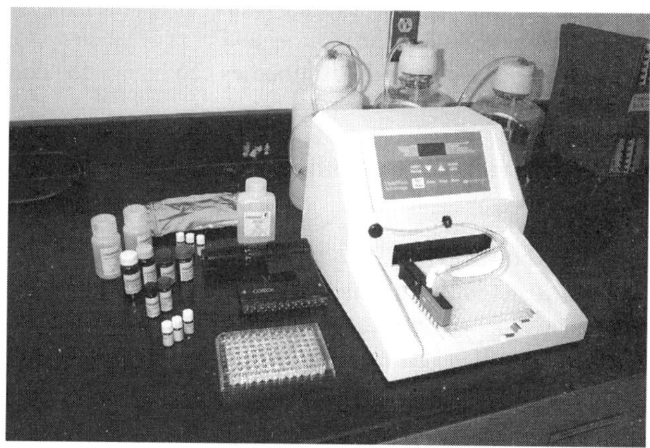

Fig. 15   ELISA Reader

One of the most useful of the immunoassays is the two-antibody "sandwich" ELISA. This assay is used to determine the antigen concentration in unknown samples. This ELISA is fast and accurate, and if a purified antigen standard is available, the assay can determine the absolute amounts of antigen in unknown samples. The sandwich ELISA requires two antibodies that bind to epitopes that do not overlap on the antigen. This can be accomplished with either two monoclonal antibodies that recognize discrete sites or one batch of affinity-purified polyclonal antibodies.

To utilize this assay, one antibody (the "capture" antibody) is purified and bound to a solid phase. Antigen is then added and allowed to complex with the bound antibody. Unbound products are then removed with a wash, and a labeled second antibody (the "detection" antibody) is allowed to bind to the antigen, thus completing the "sandwich". The assay is then quantitated by measuring the amount of labeled second antibody bound to the matrix, through the use of a colorimetric substrate. A major advantage of this technique is that the antigen does not need to be purified prior to use, also that these assays are very specific. However, one disadvantage is that not all antibodies can be used. Monoclonal antibody combinations must be qualified at "matched pairs", meaning that they can recognize separate epitopes on the antigen.

Unlike Western blots, which use precipitating substrates, ELISA procedures utilize substrates that produce soluble products. Ideally the enzyme substrates should be stable, safe and inexpensive. Popular enzymes are those, which convert a colorless substrate to a colored product, e.g. p-nitrophenylphosphate (pNPP), which is converted to the yellow p-nitro phenol by alkaline phosphatase. Substrates used with Peroxidase include 2,2¢ -azo-bis(3-ethylbenzthiazoline-6-sulfonic acid) (ABTS), o-phenylenediamine (OPD) and 3,3¢5,5¢-tetramethylbenzidine base (TMB), which yield green orange and blue colors, respectively. A table of commonly used enzyme-substrate combinations is included in Appendix A.

1. The sensitivity of the Sandwich ELISA is dependent on four factors:
2. The number of molecules of the first antibody that are bound to the solid phase.
3. The avidity of the first antibody for the antigen.
4. The avidity of the second antibody for the antigen.
5. The specific activity of the second antibody.

The amount of the capture antibody that is bound to the solid phase can be adjusted easily by dilution or concentration of the antibody solution. The avidity of the antibodies for the antigen can only be altered by substitution with other antibodies. The number and type of labeled moieties it contains determine the specific activity of the second antibody. Antibodies can be labeled conveniently with iodine, enzymes, or biotin.

General Protocol for the Sandwich ELISA method

1. Before the assay, both antibody preparations should be purified and one must be labeled.
2. For most applications, a polyvinylchloride (PVC) microtiter plate is best; however, consult manufacturer guidelines to determine the most appropriate type of plate for protein binding.
3. Bind the unlabeled antibody to the bottom of each well by adding approximately 50 mL of antibody solution to each well (20 mg/mL in PBS). PVC will bind approximately 100 ng/well ($300ng/cm^2$). The amount of antibody used will depend on the individual assay, but if maximal binding is required, use at least 1 mg/well. This is well above the capacity of the well, but the binding will occur more rapidly, and the binding solution can be saved and used again.
4. Incubate the plate overnight at 4°C to allow complete binding.
5. Wash the wells twice with PBS. A 500-ml.-squirt bottle is convenient. The antibody solution washes can be removed by flicking the plate over a suitable container.
6. Incubating with blocking buffer must saturate the remaining sites for protein binding on the microtitre plate. Fill the wells to the top with 3% BSA/PBS with 0.02% sodium azide. Incubate for 2hrs to overnight in a humid atmosphere at room temperature. (Note: Sodium Azide is an inhibitor or Horseradish Peroxidase (HRP). Do not include Sodium Azide in buffers or wash solutions, if an HRP-labeled antibody will be used for detection.)
7. Wash wells twice with PBS.
8. Add 50 mL of the antigen solution to the wells (the antigen solution should be titrated). All dilutions should be done in the blocking buffer (3% BSA/PBS with 0.02% sodium azide). Incubate for at least 2 hrs at room temperature in a humid atmosphere.
9. Wash the plate four times with PBS.
10. Add the labeled second antibody. The amount to be added can be determined in preliminary experiments. For accurate quantitation, the second antibody should be used in excess. All dilutions should be done in the blocking buffer.
11. Incubate for 2 hrs or more at room temperature in a humid atmosphere.
12. Wash with several changes of PBS.
13. Add substrate as indicated by manufacturer. After suggested incubation time has elapsed, optical densities at target wavelengths can be measured on an ELISA reader.

Note: *Some enzyme substrates are considered hazardous, due to potential carcinogenicity. Handle with care and refer to Material Safety Data Sheets for proper handling precautions.*

For quantitative results, compare signal of unknown samples against those of a standard curve. Standards must be run with each assay to ensure accuracy.

Troubleshooting ELISA Assays

- Interpret the control results.
- If the negative controls are giving positive results, there may be contamination of the substrate solution, or contamination of the enzyme-labeled antibody, or of the controls themselves.
- If no color has developed for the positive controls or for the samples, check all reagents for dating, concentration, and storage conditions. Check the integrity of the antibody reagent.
- If very little color has developed for the positive controls and the test samples, check the dilution of the enzyme-labeled antibody, and the concentration of the substrate.
- If color has developed for the test samples but not the positive or negative controls, check the source of the positive controls, their expiration date and their storage. Have they been stored in a dilute form, so that the antigen may have adhered to the surface of the storage vessel?
- If color can be seen, but the absorbance is not as high as expected, check the wavelength setting.
- When rerunning an assay while troubleshooting, change only one factor at a time.

BASIC REQUIREMENT

1. 96 well ELISA plate
2. Micropipette
3. NaHCO3 Buffer 25mM 9pH-9.6)
4. Milk Powder,
5. TBS/PBS
6. Antigen – samples from plants and animals
7. Antibodies:

    a. Primary Ab – Anti Ribulose Biphosphate Decarboxilase (RUBISCO-Plant)
    b. Secondary Ab (Ab conjugated with enzyme) – Goat anti rabbit + Horse Radish Peroxidase (HRP)

8. Spectrophotometer or Colorimeter or ELISA Reader
9. Eppendorf

- To coat the plate with the appropriate antigen, fill the micro wells of a Nunc Maxi-Sorp Immuno Plate with 100uL of the diluted antigen.
- Incubate at 4°C overnight
- Wash the unbound antigen off the plate by flicking the contents of the plate into the sink, fill the wells with DI water, flick again, repeat 2X with PBS-Triton.
- Block non-specific binding by adding 200uL of 1%BSA/PBS
- Incubate for 30-60minutes at Room Temperature (RT)
- Wash plate as above
- Add 100uL of (diluted) samples to appropriate wells. Be sure to include positive and negative controls, and, if necessary, a standard curve.
- Incubate for 1hour at RT
- Repeat washing step
- Prepare appropriate dilution of the second step antibody conjugated either with Alkaline Phosphatase or Horseradish Peroxidase. (Antibodies should be titrated for optimal dilution)
- Add 100uL of second step antibody to wells and incubate for 1hr
- Repeat washing step

- Prepare substrate solution*
- Add 100uL of substrate to well and incubate at RT for 60min
- Add stopping solution if appropriate
- Read plates on an ELISA plate reader*

*Common Substrates and the appropriate plate reader setting

- ABTS: 405-410 nm
- TMB: non-stopped 620-650 nm, stopped 450 nm
- OPD: non-stopped 450 nm, stopped 490 nm
- pNPP: 405-410 nm
- BluePhosTM: 595-650 nm

REQUIRED SOLUTIONS

PBS (Phosphate-Buffered Saline, Ca/Mg free)

Solute	1x	10x
NaCl	8g	80g
KCl	0.2g	2g
Na_2PO_4 (anhydrous)	1.15g	11.5g
KH_2PO_4 (anhydrous)	0.2g	2g

N.B.:

- $Na_2HPO_4.2H_2O$ can also be used
- add 1.42g for 1x and
- 14.2g for 10x PBS.
- Make up to 1 litre with DW

PBS wash

100ml 1x PBS
10µl Tween (0.01%, final concentration)

CITRATE /ACETATE BUFFER, $p^H6.0$

1. 500ml 0.1M Sodium Acetate (Sodium acetate - 4.1g. DW H_2O - 500ml).
2. 100ml 0.1M Citric acid (Citric acid - 2.1g. DW H_2O - 100ml).

Titrate 1 with 2 to a final pH of 6.0. Citrate/Acetate Buffer, p^H-6 can be stored frozen at -20°C.
TMB Solution - 10mg/ml in DMSO (store at 4°C).

The following details the method for use with TMB as the substrate (cover wells with clingfilm during incubations to prevent evaporation).

(1) Add 3ul H_2O_2 to 20ml of Citrate/Acetate buffer. To every 5ml of buffer add 50ul TMB solution.

N.B.: Colour change is from colourless to blue (yellow when acid is added).

(2) Coat plate with antigen stock solution (1ug/ml in PBS, 50ul/well) for at least 24hrs or over the weekend at 4°C or O/N at 37°C.

(3) Remove antigen and wash 3 times with 1x PBS.

(4) Block with 1% BSA (100ul/well) in 1x PBS for at least 1hr at 37°C.

(5) Remove BSA by flicking plate.

(6) Wash 3 times with 1x PBS containing 0.01% Tween

(7) Add anti-serum (diluted as appropriate, 1/100 etc. with 1x PBS) and incubate for 2hrs @ 37°C.

N.B.: Use positive and negative controls.

(8) Wash 4 times with 1x PBS containing 0.01% Tween.

(9) Add Peroxidase conjugate (50ul/well) diluted 1:1000 with 1% BSA (diluted in washing solution as in 8)) and incubate for 1hr @ 37°C. N.B.: Use PRAM (Peroxidase-conjugated Rabbit Anti-Mouse) for Ab raised in mice and PSAR (Peroxidase-conjugated Swine Anti Rabbit) for Ab raised in rabbits.

(10) Wash 4 times with 1x PBS containing 0.01% Tween.

(11) Add substrate (100ul/well).

(12) Incubate for 1-5 mins in the dark at RT. Check reaction for colour change. Wells should turn blue.

(13) Stop reaction by the addition of 25ul 0.1M H_2SO_4/well.

(14) Read OD @ 450nm for TMB.

15

ELISA PROCEDURE

PROCEDURE

A. Take different organs of animals and grind them with PBS and can use them as different antigens.

B. Fixation of antigen to the well:

- Take 96 well ELISA plate and wash with distilled water carefully.
- Wash with PBS twice and discard forcefully.
- Add 100 μl of NaHCO3 Buffer to each well of 1st. horizontal rows and three vertical rows.
- Add serial dilution of antigen (1), (2) and (3) to the vertical rows, as shown (Fig)
- Next to the horizontal rows (A) and (B) add 20 μl of any antigen of which the titre has to be calculated.
- Cover the plate with lid and leave it undisturbed at room temperature overnight. The antigen will be attached to the sides.

Addition of Antibodies (Primary and Secondary)

1. Discard excess antigen forcefully form the wells.
2. To block the gaps left after antigen binding, add 100 μl of 3% milk solution to each well. Milk solution is prepared by adding 3gm. of milk powder in 100 ml of PBS.
3. Incubate for 10 mins. At room temp.
4. Meanwhile prepare a milk –Ab solution as follows:
   1% milk solution in PBS + Primary Ab (RUBISCO) at 100:1 ratio, i.e. add 10μl of Ab in 1ml or 1000 μl of 1% milk in TBS (Tris Buffer Saline)
5. Now discard the solution forcefully from the well kept for incubation.
6. Add milk solution containing Ab to all the wells and incubate for 10mins. at room temperature or 37°C.
7. Meanwhile prepare serial dilution of Secondary Ab (Goat anti rabbit with HRP) in another ELISA plate as follows:

   (a) Add 100 μl of 1% milk solution in PBS or TBS to each well of horizontal rows.
   (b) Add 50 μl of concentrated Ab conjugated with enzyme (Secondary Ab)
       to the first well mix well and transfer 50 μl from it to the 2nd. And go on diluting the wells in the same manner. Discard the last 50 μl.
   (c) This serially diluted Abs can be used for addition in 2nd. Horizontal rows.

8. Also prepare a constant concentration of Ab with enzyme and add to it 5 μl of 1% milk solution in TBS. This constant concentration of Ab can be used for 3 vertical rows.

9. After the incubation time is over, discard the solution in the wells forcefully by tapping the plate.
10. Wash the wells with 100 μl of PBS or TBS twice and discard TBS forcefully each time.
11. Add 100μl serially diluted secondary Ab to the two horizontal rows, as shown: (Fig.)
12. Then add 100 μl of (I: 2000) diluted Ab-milk solution to each of vertical rows.
13. Incubate the plate for 20 mins. at 37°C.
14. Discard and wash twice with PBS or TBS.
15. Add 100 μl substrate to each well (Substrate + Chromogen, called Trimethyl Benzedene or TMB + H_2O_2).
16. By adding substrate the blue colour appear with varying intensity in each well. Incubate at 37°C for 10 mins.
17. Read the absorbance using red filter at 620nm. If reaction has to be stopped mix 0.1N H_2SO_4 to each well. Read at 495nm.
18. Plot a graph of absorbance against number of wells.

OBSERVATION

Take 100 μl of working substrate and dilute with 1ml DW. Use this as blank, i.e. set zero or 100% absorbance with this blank. Measure the OD of all the wells and the value that will give the max OD will be the titre value.

16

WESTERN BLOT ANALYSIS

AIM

To perform western blot analysis

PRINCIPLE

Identification of a specific protein in a complex mixture of protein or Ab to a given protein can be accomplished by western blot. In this technique protein is electrophoretically separated on a polyacrylamide slab gel. The protein bands are transferred into a nitrocellulose membrane by electrophoresis or by diffusion such that the membrane gets a replica of the gel containing protein bands. Flooding the membrane with primary Ab identifies the individual proteins. After this the Ag – Ab can be detected by adding a secondary Ab conjugated with enzyme. The band is visualised by adding substrate, which being a Chromogen on conversion to product gives colour. The intensity of colour shows the quantity of Ag – Ab complex formed.

Sufficiently separated proteins in an SDS-PAGE can be transferred to a solid membrane for WB analysis. For this procedure, an electric current is applied to the gel so that the separated proteins transfer through the gel and onto the membrane in the same pattern as they separate on the SDS-PAGE. All sites on the membrane, which do not contain blotted protein from the gel, can then be non-specifically "blocked" so that antibody (serum) will not non-specifically bind to them, causing a false positive result.

To detect the antigen blotted on the membrane, a primary antibody (serum) is added at an appropriate dilution and incubated with the membrane. If there are any antibodies present which are directed against one or more of the blotted antigens, those antibodies will bind to the protein(s) while other antibodies will be washed away at the end of the incubation. In order to detect the antibodies, which have bound, anti-immunoglobulin antibodies coupled to a reporter group such as the enzyme alkaline phosphatase are added (e.g. Goat anti-human IgG- alkaline phosphatase). This anti-Ig-enzyme is commonly called a "second antibody" or "conjugate". Finally after excess second antibody is washed free of the blot, a substrate is added which will precipitate upon reaction with the conjugate resulting in a visible band where the primary antibody bound to the protein.

Enzymes and substrates for western blot analysis: A number of enzymes and substrates have been employed for western blot developing, like alkaline phosphatase, Horseradish Peroxidase, NBT, BCIP.

1. Nitroblue Tetrazolium (NBT): prepared in buffer

    NBT = 3.0mg

    Buffer : Tris-HCl = 0.1M (pH 8.8)

$$NaCl \quad = 0.1M$$
$$MgCl_2 \quad = 0.005M$$

2. 5-Bromo-4-Chloro Indolyl Phosphate (BCIP): .

BCIP	= 1.5mg
Dimethyl Sulphoxide	= 10µl
2M Tris-HCl (pH9.8)	= 0.25ml

Mix solution 1 and 2 and make up to 10ml

3. Horseradish Peroxidase (HRP):

    A. 30mg of 4-Chloro-1-Naphthol in 10ml Methanol

    B. 30µl of H_2O_2 in 40ml TBS

Mix solution A and B at room temperature.

Mechanism of Transfer of protein from gel to membrane: Transfer of protein bands from an Acrylamide gel onto a membrane by diffusion by the capillary action of buffer is called 'Capillary Blotting'. For this a number of solutions are required.

1. Urea Buffer:

Sodium Chloride	=	0.29g
EDTA	=	0.07g
Tris	=	0.12g
Urea	=	24.00g

Adjust the pH to 7.0 and make the volume to 100ml with distilled water

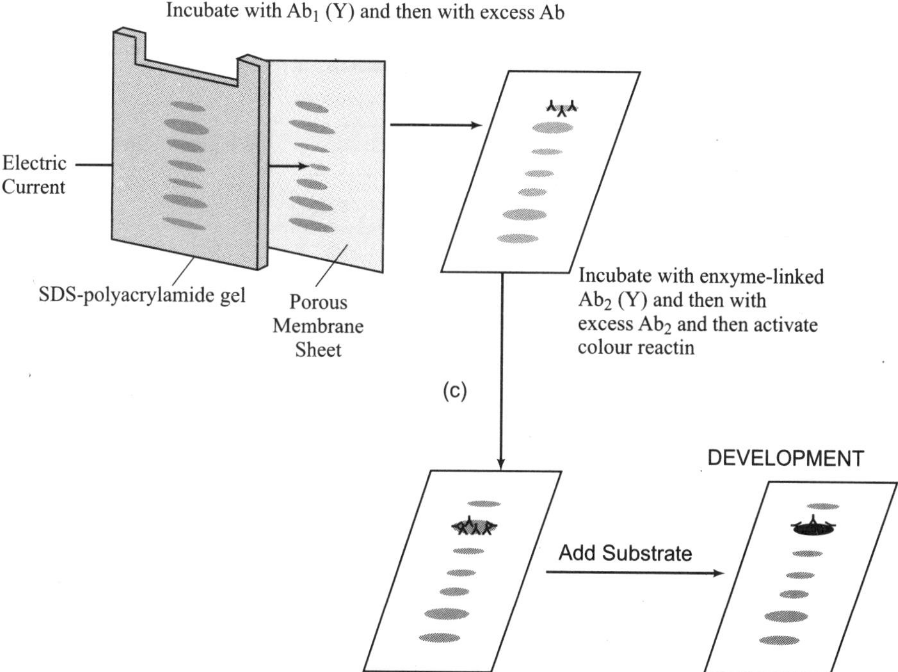

Fig. 16   Technique of Western Blotting

2. Transfer Buffer:

Sodium Chloride	=	2.9g
EDTA	=	0.7g
Tris	=	1.2g

3. Phosphate Buffer saline (PBS- pH7.2)

Sodium Chloride	=	8.00g
Potassium chloride	=	0.2g
Disodium Hydrogen Phosphate	=	1.15g
Potassium dihydrogen Phosphate	=	0.20g
Distilled water	=	1 litre

PROCEDURE

1. Remove SDS from the gel by washing the gel with urea buffer for 30 min.
2. Measure the size of the gel and cut 6 sheets of Whatman No.3 paper and 2 sheets of nitrocellulose membrane.
3. Assemble the sandwich as given in fig.:

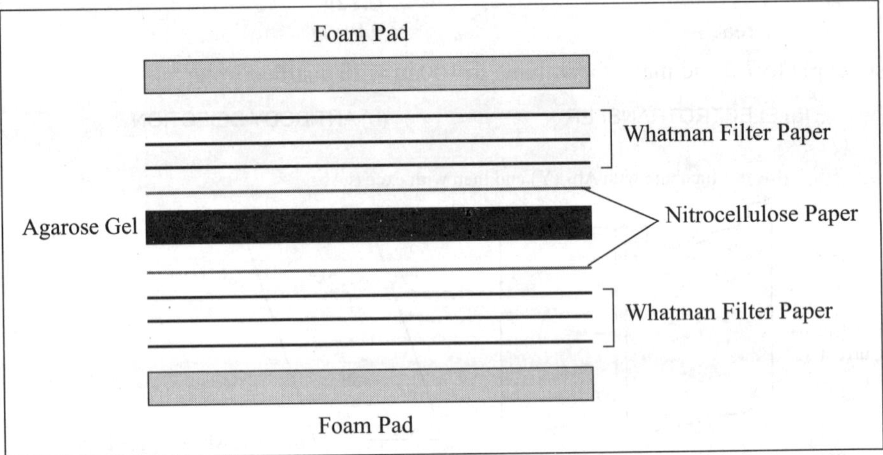

Fig. 17 Components of a Sandwich

4. Put rubber bands to keep the sandwich intact and secured.
5. Keep the sandwich in transfer buffer kept in a tray.
6. Keep a lightweight on top of the sandwich to prevent it from floating and to ascertain that it is kept immersed completely. Avoid trapping any air bubble while preparing the sandwich assembly.
7. Keep for transfer for 12 to 18 hrs.
8. Remove nitrocellulose membrane and proceed for Ab labelling or can be stored in a box for more than 6 months.

ELECTROPHORETIC TRANSFER

1. Soak the gel, Whatman paper and nitrocellulose (NC) transfer buffer (80% running buffer + 20% methanol)
2. Transfer protein electrophoretically using the transfer apparatus.

PROCEDURE FOR DEVELOPING WESTERN BLOT

1. Take the NC paper and shake in the blocking agent for 1 – 2 hrs. (3% milk in TBS, 10ml Tris-HCl, and pH 7.3, contains 0.15M NaCl).
2. Transfer this paper to 1% primary Ab solution made in blocking agent. Keep at room temperature for 1 hr. with constant shaking.
3. Rinse the paper in distilled water and keep in another container with TBS for 10min. at room temperature.
4. Repeat step 3 with fresh TBS.
5. Add secondary Ab conjugated with alkaline phosphatase (1:2000 dilutions) in 1% milk in TBS.
6. Shake for 1hr. at room temperature.
7. Rinse in distilled water and wash in TBS 10 min. twice.
8. Add the substrate solution (NBT + BCIP) in dark.
9. Colour develops within ten min. Take a photograph immediately.

17

PURIFICATION OF IMMUNOGLOBULIN

AIM

To purify IgG Antibody from the given serum

PRINCIPLE

Proteins are charged molecules and are associated with solvent molecules in solution. When this association is disturbed by addition of inorganic salts, like Ammonium sulphate, molecules tend to aggregate and precipitate out of the solution as this neutral salt affects the balance between electrostatic forces tending to keep the proteins in solution and hydrophobic forces that cause proteins to aggregate and precipitate in aqueous solution.

Ammonium sulphate precipitates protein very efficiently as it has very high soluation energy and hence attracts the solvent molecules towards it. Ammonium sulphate is non-toxic and stabilizes protein in solution.

PROCEDURE

I. Ammonium Sulphate Fractionation

1. Take serum in a conical flask.
2. Add 100% Ammonium Sulphate drop wise with constant mixing to 33.3% saturation in ice.
3. Keep it ice for 20 min.
4. Centrifuge it at 5000rpm for 5min.
5. Suspend the pellet in PBS and let be dialyzed overnight in phosphate buffer.

II. Pre Treatment of Dialysis Bag

1. Boil the dialysis bag in distilled water (DW) for 10 min.
2. Boil it for 30min. in 0.2M $NaHCO_3$.
3. Wash this bag in DW.
4. Boil it in 1mM. EDTA solution.
5. Wash the bag in DW and store it in 50% alcohol.

III. Affinity Chromatography

1. Suspend affinity chromatography material (Cyanogen activated Sepharose 4B) in phosphate buffer.
2. Wash five times with ImM HCl.

3. Wash with coupling buffer (0.1M NaHCO$_3$ pH-8.3, and 0.5MnaCl) to remove HCl.
4. Dissolve the protein (antigen, IgG) to be coupled in coupling buffer and take O.D. at 260 – 280nm.
5. Mix the protein and the above treated gel (Sepharose 4B) and keep for 2hrs. at room temperature.
6. Transfer these then to blocking reagent that is 0.02 Glycine (pH8) and keep for 2hrs.
7. Wash it with coupling buffer and then with acetate buffer (0.1M Sodium Acetate, pH-4, and 0.5M NaCl).
8. Finally wash it with coupling buffer and store it overnight at 4°C.
9. Pack the beads in the column and layer the IgG fraction in PBS on the top of the column.
10. Wash the column twice with PBS.
11. Elute with 0.02M Glycine (pH-8) and thrice with acetate buffer (0.1M Sodium Acetate).
12. Elute with acetate buffer at different pH to obtain different Ig
    - pH-6.0 to obtain IgG1
    - pH 4.5 to obtain IgG2a
    - pH 3.5 to obtain IgG2b
13. Take O.D. at 280 and 260nm of all the fractions and calculate the total amount of IgG purified.
    Calculation: $1.55 \times$ O.D. $280 - 0.76 \times$ O.D. $260 =$ mg protein / ml

18

CONJUGATION OF ENZYME WITH ANTIBODY

AIM

Coupling of Alkaline Phosphatase to an Antibody

PRINCIPLE

Different enzymes like Alkaline Phosphatase and Horse Radish Peroxidase can be coupled to antibody. This helps in the detection and quantification of antigen by a coloured reaction, which arises upon addition of appropriate substrate. Enzymes are coupled to Fc region of Ab by means of a reaction that involves gluteraldehyde. Such antibodies are used in assays like ELISA and Western Blot analysis.

PROCEDURE:

1. Dialyse alkaline phosphatase against phosphate buffer (0.1N).
2. Adjust the final concentration to 5mg/ml.
3. Dialyze the Ab also against phosphate buffer.
4. Adjust the concentration to 10mg/ml.
5. Take 0.5ml of Ab and add 0.5ml of enzyme.
6. Add 50μl of 1% gluteraldehyde to the above solution.
7. Mix thoroughly and incubate at room temperature for 4 – 6hrs. or overnight.
8. Add 5μl of 1M-ethanolamine to the above solution and incubate for 2hrs.
9. Pass this through 10-bed volume of Sephadex G.25 in PBS and ethanolamine.
10. Collect half volume fraction with PBS and Ethanolamine. Check 2μl of fraction with 50μl of substrate.
11. Pool the fractions, which give colour and store at 4°C. in 50% glycerol.

19

PREPARATION OF LYMPHOCYTES

AIM

Isolation of mononuclear cells from peripheral blood on ficoll gradient

PRINCIPLE

The separation of RBC and the immunological cells, i.e. lymphocytes, is based on the differences in density with reference to gradient. In ficoll gradient RBCs settle down and the lymphocytes are found in the interphase. This forms the basis of separation of lymphocytes from the blood samples.

PROCEDURE

1. Collect 10ml of venous blood in 2ml of Alsever's solution (2ml).
2. Take 3ml of blood and add it to a beaker containing Alsever's solution.
3. Measure volume and add PBS in a tube and make up volume to 10ml and mix well.
4. Take 4ml of ficoll in another graduated tube.
5. Add the blood slowly and carefully onto this, so that it forms a red layer over the transparent layer of ficoll, till total volume is 25ml.
6. Centrifuge at 750rpm for 25min.
7. Four layers will be formed:

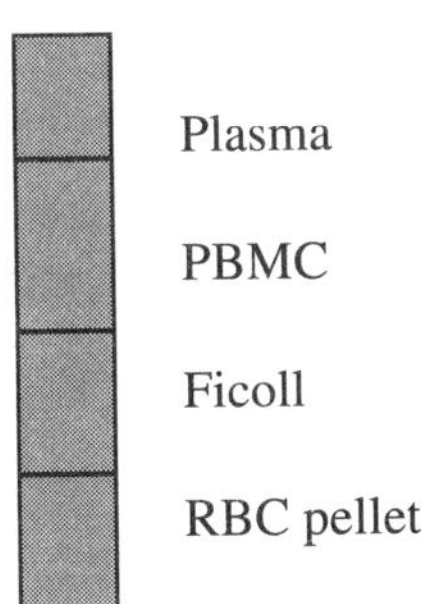

Fig. 18 Ficol gradient.

8. Remove PBMC layer.
9. Add PBS and centrifuge at 1500 rpm for 10 min. repeat thrice.
10. Add to 50µl of above preparation, 50µl of 0.4% Trypan Blue.
11. Count the viable cells after loading the sample on a Haemocytometer. Dead cells pickup the dye.

20

PLAQUE FORMING CELL ASSAY

AIM

Performing plaque forming cell assay for in-vitro antibody production. Diffusion, Electrophoresis, ELISA, and Western Blotting etc.

This technique can be used for determining the number of antibody producing cells. When spleen cells collected from mouse are mixed with antigen (sheep/goat RBC), it forms a clear zone in the Cunningham Szenberg Chamber. This clear zone is called 'plaque'. The principle is as follows:

"The spleen cells contain antibody producing cells. These antibodies on binding to sheep/goat RBC (Ag) enable binding of complement, which leads to lysis. Hence, the zone around antibody producing cells will become clear forming a plaque, which will give us the total number of antibody producing cells".

REQUIREMENT

1. Micro glass slides
2. Double adhesive tape,
3. PBS
4. RRBC
5. BSA
6. Guniea pig complement
7. Minimum Essential Medium (MEM)
8. Syringe
9. Eppendorf
10. Wax
11. Complement (Serum other than Rabbit)

PROCEDURE

A. Collection of Spleen Cells

1. Inject 100µl of 5% SRBC/GRBC/RRBC to the mice
2. After 4 days dissect the mice and take out the spleen
3. Crush the spleen with 1ml. of PBS
4. The suspended spleen cells are now ready for use.

B. Preparation of Slide Chamber

1. Soak the microslids for 1/2h in $3NHNO_3$
2. Rinse with distilled water and dry
3. Take two glass slides. To one attach the tape at 3 places – on two corners and one at the middle. Place the second slide and press on top of the first.
4. Now create dual chamber using 3 pieces of double adhesive tape yielding a final volume of 100 µl per chamber

Immunization Technique

1. Immunize mice with 200 µl of 10% SRBC/GRBC/RRBC in 0.95 NaCl injected intraperitonially.
2. Sacrifice the animal on day 5 and remove spleen
3. Press the spleen through wire mesh to separate the spleen cells

PFC development

Prepare reaction mixture using the following composition:
- 25 µl SRBC/GRBC/RRBC-2.5%
- 25 µl guinea pig complement
- 100 µl spleen suspension
- 5% bovine serum
- 100 µl MEM

Dilute all reagents with MEM to require concentrations

Plaque counting

Microslides can be easily conted visually under low power microscope. Values of $PFC/10^3$ spleen cells can be calculated using the following formula:

$$\frac{\text{Total volume of reaction mixture}}{\text{Chamber volume}} \times \text{dilution factor} \times PFC/\text{cell}$$

Note: Optimal number of plaques is in the range 30-50/slide

Represent the result in the following tabulated form:

Replicate Number	$PFC/10^6$ spleen cells
1	
2	
3	
4	
5	
6	
7	

Plaque Assay Method-2

Materials

- Grace's Insect Medium, 2X (e.g. GIBCO # 11667)
- Fetal Bovine Serum (Heat Inactivated), (e.g. GIBCO # 16140)
- 3% low-melting agarose in ddH_2O

- Sterile water
- 50 ml sterile conical screw-top tubes
- 37°C water bath
- microwave

Protocol

Prepare infected monolayer of cells

1. Grow a suspension culture of Sf9 cells to a density of less than $3 \times 10E6$.
2. Dilute this culture to to a density of 5 to $6 \times 10E5$.
3. For a 6-well culture dish, transfer 2 ml of this cell suspension to each well. For 6 cm dishes, double all volumes in this protocol. Scale by surface area.
4. This cell number will depend somewhat on the cell line and can be adjusted up or down according to your results.
5. If you don't have a confluent monolayer by day 3, increase the cell number. There should be space available on Day 2.
6. Let the cells settle for at least 30 min. Longer is just fine. Look at them and make sure they're stuck.
7. Meanwhile, dilute your virus to 1 ml aliquots at dilutions of 10E-4, 10E-5, 10E-6 and 10E-7.
8. It is critical to use fresh tips when transfering each serial dilution to the next tube or you will overestimate titer.
9. After your cells have attached well to the plates, tip them up a bit to drain medium to one side, and aspirate off the media. It's not critical to remove all the medium at this point.
10. Quickly add 1 ml of diluted virus to each well of the 6-well plate. If you're using 6 cm dishes, don't try to do too many plates at once. It's easy to dry the cells and kill them. If you do this, you'll see a characteristic arc of dead cells on the edge of the plate that was tipped up to drain off medium.
11. Transfer the plates to a rocking platform and slowly rock for at least a couple hours, four hours is probably better, though after this the benefit diminishes substantially.
12. Prepare your overlay agarose just before use.
13. This is a bit tricky and the timing is critical. Low melt agarose will solidify at temperatures near or slightly below 37°C, but it's a bit easier on the cells if you can manage this.

Mix the reaction mixture as follows:

1. 1 part 2X Grace's Medium supplemented with 20% Fetal Calf Serum. Ignore the dilution due to FCS.
2. 1 part 3% Sea Plaque Agarose in ddH_2O.
3. Melt the agarose completely in a microwave being careful to allow air to escape to avoid exploding glassware.
4. Allow the agarose to cool slightly, to near 70°C and then aliquot this 20 ml to each 50 ml conical tube.
5. Add 20 ml of room temp or warmer 2X Graces/FCS to each 20 ml aliquot of agarose, then place in a 35-37°C water bath.
6. Remove the overlay agarose from the water bath one tube at a time and check the temperature. Let it cool to at least 38°C, but preferably less than 37°C.

Overlay agarose onto infected cell monolayer

1. When ready, tip up the plates and let drain to one side. Aspirate all the medium off, and I usually go back a second time and aspirate each well a second time at the edge to get all medium out.
2. Return the plate to level and quickly add approximately 3 ml of molten overlay mix to each well by allowing it to slide down the far wall of the well and onto the plate.
3. This is the critical stage.
4. If your are too slow, your cells will dry out and you will see the arc of dead cells that I mentioned above on the high side of the plate.
5. If you leave too much medium on the plate, they will be sloppy and the plaques will spread or smear and won't be well defined.
6. If you put the overlay on too hot, you'll cook the cells especially at the point where the overlay first touches the plate.
7. If you tip the plate up on the near edge to drain away from you, and you slide the agarose down the far side of the well, you will be able to see easily which factors you need to adjust as different sides of the plate will be affected.
8. After overlaying the cells let the plates sit level in the hood for 30 or so minutes to dry a bit and solidify.
9. If your plaques are smearing, it is a good idea to crack open the lids for 20 minutes after they're poured to allow more moisture to escape.
10. Place in a 27°C, 98% humidity controlled incubator for at least 3 days. If you don't have a humidity controlled incubator, the plates can be sealed in a plastic bag or tupperware-type container with damp towels in the bottom, but this invites mold, and may limit oxygen somewhat.

Staining the Plates

1. Prepare a solution of 1% Neutral red. Ten mls of this will stain about 1000 assays.
2. Prepare an overlay agarose solution as above, but only prepare 1 ml for each assay.
3. Add 1/100th volume of the 1% Neutral Red solution to the molten agarose (e.g. 100 μ liters to 10 ml).
4. Add approximately 1 ml of the Red Agarose to each well of a 6-well dish. Be sure the plate is level until the agarose sets up.
5. You only need to add enough Red Agarose to cover the surface evenly.
6. Return the plate to the incubator for at least 4 hours. After a couple hours the plaques will begin to appear as clear spots among stained cells.
7. Leave the plate overnight before counting. This makes it easier. If plaques aren't developing, and this depends somewhat on you cell line, leave the plates 4 or 5 days before staining, but this isn't usually necessary.
8. Do some controls and verify that longer incubations don't give higher titer results with your medium and cells.

MICROBIOLOGY

LABORATORY SAFETY AND BASIC REQUIREMENTS

Just like any other activity in a laboratory, all microbiology practicals require the user to adopt good laboratory practice. These brief notes and hints are intended to help those involved in the various activities to carry them out safely. Remember that before any practical activity is undertaken a risk assessment should be performed to ensure there is minimal hazard to all concerned. If there is any doubt about the assessment of the risk, reference must be made to safety texts or expert advice taken.

MICROBIOLOGICAL SAFETY

The techniques and activities described here and the micro-organisms suggested present minimum risk given good practice. It is therefore essential that good microbiology laboratory practice is observed at all times when working with any microbes.

There are five areas for consideration when embarking on practical microbiology investigations which make planning ahead essential.

1. Preparation and sterilisation of equipment and culture media.
2. Preparation of microbial cultures as stock culture for future investigations and inoculum for current investigation.
3. Inoculation of the media with the prepared culture.
4. Incubation of cultures and sampling during growth.
5. Sterilisation and safe disposal of all cultures and decontamination of all contaminated equipment.

Good organisational skills and a disciplined approach ensure that every activity is performed both safely and successfully.

PROTECTION

Food or drink should not be stored or consumed in a laboratory that is used for microbiology. One should not lick labels, apply cosmetics, chew gum, suck pens or pencils or smoke in the laboratory. Hands should be washed with disinfectant soap after handling microbial cultures and whenever leaving the laboratory. If hand contamination is suspected, then the hands should be washed immediately with disinfectant soap. To ensure that any wounds, cuts or abrasions do not get infected or infection is passed on, protect them by the use of waterproof dressings or wear disposable surgical gloves.

GENERAL PERSONAL SAFETY

Each individual embarking on these, or any other, activities is responsible for their own safety and also for the safety of others affected by their work (students, technicians, and teachers). The individual must include in the planning and performance of the investigation a risk assessment to assess any hazard that the investigation may pose and ways of minimising it. Points to consider include: safe storage and culturing of micro-organisms; emergency procedures such as dealing with spillages and disposal of all contaminated material. No one should perform any microbiological procedures without receiving appropriate training from a competent person. To minimise the chance of contamination of the user, any other individual, the environment or the microbial culture good laboratory practice is required. GLP requires us to consider all cultures as potentially pathogenic.

ASEPTIC TECHNIQUE

Sterile equipment and media should be used to transfer and culture micro-organisms. Aseptic technique should be observed whenever micro-organisms are transferred from one container to another. Contaminated equipment should preferably be heat sterilised by either incineration or autoclaving. A suitable chemical disinfectant can be used but this may not ensure complete sterilisation.

ELECTRICAL SAFETY

Some microbiology investigations use bioreactors that require oxygenation and this is usually supplied by the use of an aquarium air pump. Care should be taken to ensure that no liquid comes into contact with electrical mains power. The same care should apply if a magnetic stirrer is to be used to mix the growth medium in a bioreactor.

BASIC REQUIREMENTS

A. Instruments and Appliances

- Refrigerator
- Incubator
- Autoclave/Pressure Cooker
- Microscope with oil immersion lens
- Hot plate
- Balance
- Inoculation Chamber
- Colorimeter/Spectrophotometer

B. Tools

- Inoculation needle/loop
- Steel needles
- Forceps
- Scissors
- Micrometer (ocular & stage)
- Haemocytometer

C. Glass wares

- Petriplates
- Conical flasks
- Culture tubes (rimless test tubes)
- Beaker
- Funnel
- Micro Pipette (fixed volume and variable)
- Slides & Cover slips
- Glass rod/spreader
- Graduated cylinder
- Dropper/reagent bottle

D. Miscellaneous

- Test tube rack
- Cotton (non absorbent type)
- Glass marking pencil
- Distilled water facility
- Disinfectant
- Blotting paper

21

PREPARATION OF LIQUID CULTURE MEDIA

AIM

Preparation of basic liquid media (Nutrient Broth) for routine culture of Bacteria.

THEORY

Nutrient preparations on which microorganisms can be cultures in laboratory are called 'Culture Medium'. Liquid culture medium is also called broth. One of the common types of broth used for routine cultivation of bacteria is 'Nutrient Broth'. It has only peptone and beef extract as its constituents. Peptone is a semi-digested protein, which can supply amino acids to microorganisms. Beef extract acts as a source of organic carbon, nitrogen, vitamins and inorganic salts.

REQUIREMENTS

Composition of Nutrient Broth

- Peptone = 5.0 g
- Beef Extract = 3.0 g
- Distilled water = 1000 ml
- pH = 7.0

PROCEDURE

1. Put the weighed amount of Peptone (5 g) and Beef Extract (3 g) in 500ml of distilled water in a beaker.
2. Heat with agitation to dissolve the constituents.
3. Add distilled water to make the volume 1litre.
4. Check pH (7.0). Adjust the pH if necessary by adding 1NHCl or 1N NaOH drop by drop.
5. Transfer the contents to a conical flask and plug the flask.
6. Sterilize the conical flask with autoclave/pressure cooker for 15 min. Cooled sterilized broth is now ready for routine culture of bacteria.

PRECAUTION

(a) Plugging of conical flask should be done carefully.
(b) Sterilized medium should be stored in refrigerator for further use in the laboratory.

NOTE

Students can be asked to prepare SOLID MEDIA, called 'Nutrient Agar Medium' composed of:

- Peptone   =   5 g
- Beef Extract   =   3 g
- Distilled Water   =   1000 ml
- Agar   =   15 g

22

PREPARATION OF SOLID CULTURE MEDIA

AIM

Preparation of standard solid culture media for routine culture of Bacteria

THEORY

A large number of different culture media have been developed to meet the specific requirements for culturing different bacteria and purposes. In the present experiment standard solid medium containing agar and essential ingredients, for supporting a wide variety of different bacteria, is prepared.

Composition of Standard Medium

Ingredients	Quantity
A. Basal Medium	
1. 10% Glucose sterile solution	20 ml
2. Dipotassium Hydrogen Phosphate, K_2HPO_4	7.0 g
3. Potassium Dihydrogen Phosphate, KH_2PO_4	3.0 g
4. Sodium Citrate, $Na_3C_6H_5O_7, 2H_2O$	0.5 g
5. Magnesium Sulphate, $MgSO_4, 7H_2O$	0.1 g
6. Ammonium Sulphate, $(NH_4)_2SO_4$	1.0 g
7. Agar	20 g
8. Distilled water	1000 ml
B. Trace Elements	
1. Ferrous Sulphate, $FeSO_4, 7H_2O$	0.5 g
2. Zinc Sulphate, $ZnSO_4, 7H_2O$	0.5 g
3. Manganese Sulphate, $MnSO_4, 3H_2O$	0.5 g
4. Sulphuric Acid, $H_2SO_4, 0,1N$	10 ml
5. Distilled Water	1000 ml

NOTE

- Glucose is partly degraded when autoclaved in the presence of phosphate. Hence, it is added after the remainder of the medium is sterilized.

- 5ml of the trace element solution and 1% Calcium Chloride ($CaCl_2$) is added per 1000ml of medium to supplement the essential minerals other than those present in the basal medium.
- pH should be adjusted to 7.

PROCESSING OF THE DIRECT MEDIUM

1. Specified volume of distilled water (sterilized) should be taken into bottles containing dried medium. The bottle should be capped and the content should be mixed thoroughly and vigorously, preferably in a shaker.
2. The bottle should be autoclaved for 20 – 25 min. at a pressure of 15lb/inch. Later cool to about 50°C.
3. Unscrew the cap of the bottle and later flame the mouth of the bottle for a few seconds.
4. About 15 – 20ml. of molten medium should be poured slowly onto the base of the sterile petriplate by lifting the lid of the petriplate only as much as necessary for accommodating the bottle-mouth. The molten medium should not be poured on the edges of the petriplate. Such petriplates should be discarded.
5. The petriplates, containing the medium, should be dried in incubator set at 37°C overnight with the base of the plate resting on the lids.

23

ISOLATION OF BACTERIA INTO PURE CULTURE

Three different methods for isolation of pure culture is mentioned below:

1. STREAK-PLATE METHOD

It is essentially a method for spreading bacterial cells on a sterile agar plate so that individual cells can grow into distinct colonies.

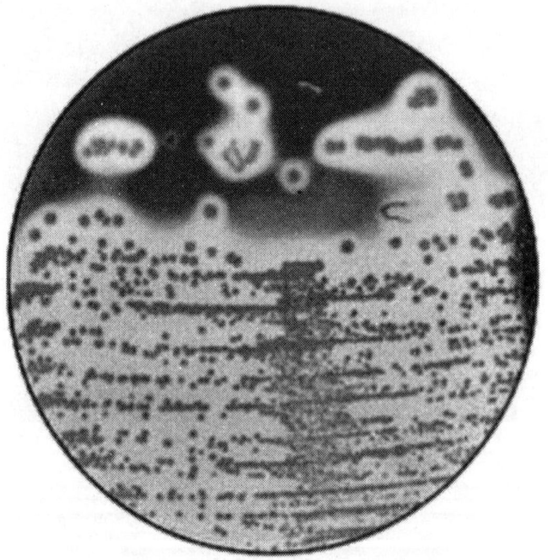

Fig. 19   Bacterial Culture in a culture plate

REQUIREMENTS

- Source of bacteria (sewage water, soil, pus, sputum, etc.)
- McCartney Bottle (screw capped glass bottles)
- Petriplates
- Burner
- Medium (nutrient agar)
- Incubator
- Culture Hood
- Wire Loop

PROCEDURE

1. The wire loop should be sterilized in burner until the loop is red hot. It should be allowed to cool for 10 – 15 seconds before use.
2. The liquid / solid sample should be collected by dipping / touching the wire loop in the liquid or solid sample.
3. Smear the sample over the 'X' marked surface of the medium by raising the lid of the petriplate (only as far as necessary), so that a well-developed inoculum is formed.
4. The wire loop should be resterilized and then 3–4 parallel lines should be streaked over the fresh surface of the same plate, as shown in the fig.
5. The base of the plate indicating index number, date, and other particulars should be labeled.
6. The plate should be incubated upside down.
7. The process should be repeated using one of these individual colonies. Pure colonies can be re-plated in separate sterile plates for obtaining pure cultures.

1. STREAK PLATE TECHNIQUE

The streak plate method is a rapid and simple technique of mechanically diluting a relatively large concentration of microorganisms to a small, scattered population of cells. The goal is to obtain isolated colonies on a large part of the agar surface, so that desired species can then be brought into pure culture. Proper streaking of plates is an indispensable tool in microbiology. In most cases a closed inoculating loop is used for streaking plates. The wire loop should not be badly oxidized or pitted or it will fail to dilute the inoculum and will scratch the surface of the agar. Streak plates can be made from a broth culture, an agar slant or from an agar plate. It is sometimes convenient to suspend a bit of growth from a solid surface in sterile saline and use this as a source of inoculum. Resuspension of colonies or cultures grown on solid surfaces dilutes the culture and makes streak plating easier. A loopful of inoculum is transferred from the source and put on the agar surface. When using a large inoculum (a turbid culture or growth from a solid surface), a small spot is spread during the initial transfer. If the inoculum is from a lightly turbid suspension, the first phase of the streaking pattern is begun. Several basic patterns are illustrated in Figure. The three-phase streaking pattern is recommended for beginners because it is most likely to give satisfactory results with suspensions having a wide range of microbial density.

There are a number of different methods for mechanically diluting microbes on a streak plate. The most common method is spreading microbes across a plate as shown in the first four figures. As the concentration of microbes increases so do the number of phases. Irrespective of the number of phases, loop is flamed between each one. The fifth plate shows an alternative method, where the streaks are not continuous, but are a series of parallel lines.

Choosing a streaking pattern is a matter of individual preference and depends upon the number of microorganisms in the sample. The fig. demonstrates the most common patterns, but they are not the only methods. The object of any streaking pattern is the continuous dilution of the inoculum to give many well isolated colonies. For multi-phase streaking it is crucial to flame the loop before starting the next phase. Note the slight overlap into the previous phase to pick up a small inoculum. To streak a plate...

STREAK FOR ISOLATION

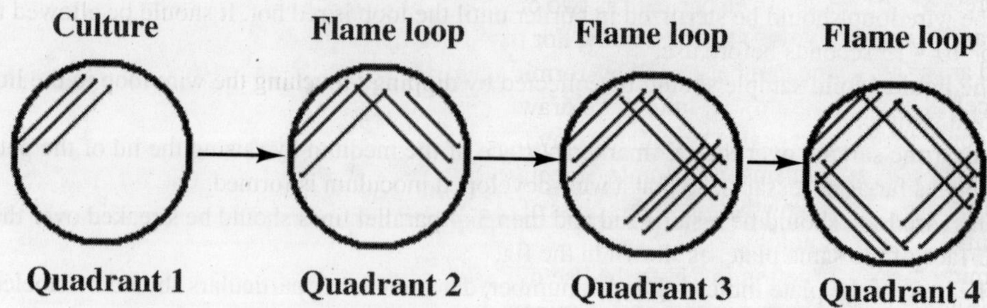

Fig. 20   A number of methods are there for streaking the culture disc

2. POUR-PLATE METHOD

A practical and common laboratory technique used in isolating pure cultures or enumerating the living microorganisms in water, milk, foods, and other materials is the pour plate technique.

To aseptically transfer liquid into a pour plate, raise one side of a Petri plate lid only just enough to allow access of the sample (from a tube or pipette). Transfer a known amount of the sample to the dish and cover immediately with the lid. Then pour 15-20 ml of sterile agar culture medium which has been melted and cooled to 45-50°C into the plate as shown in Figure. The inoculum and medium are mixed by gentle rotation ten times in one direction and ten times in the other direction. The agar must be allowed to solidify completely before the plates are inverted for incubation. After incubation both surface and subsurface colonies will be observed.

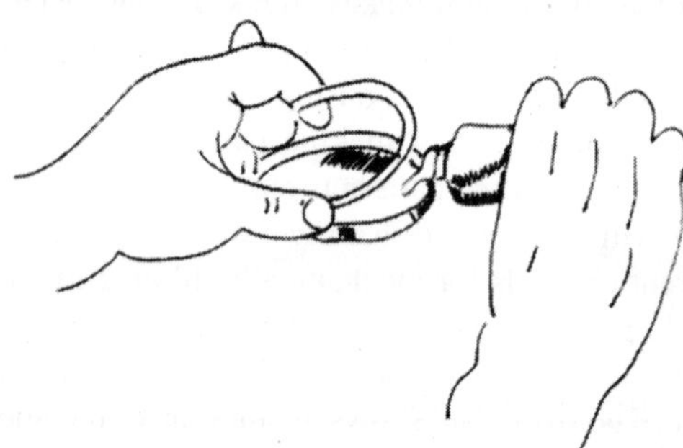

Fig. 21   The Pour Plate Technique

Pour plates allow the addition of larger amounts of liquid (1-5 ml) to an agar dish. The sample is added to the bottom of a sterile Petri plate. Molten agar is then added to the plate aseptically. It is important to only open the cover enough to allow the pouring of the agar. This prevent contamination from the environment

3. SPREAD – PLATE METHOD

Incubation is another essential technique for working with microorganisms.This method suffers from some problems. First, only those organisms which can grow on the medium, and at the temperature and atmospheric conditions of incubation, will divide and develop into colonies. Second, each colony may not represent the progeny from one cell, as two or more cells (those in clusters, chains or otherwise close to one another) can give rise to one colony. For these reasons the counts obtained from the plate-count method are given as the number of colony-forming units (CFU's) per ml (or gram) rather than the number of cells per ml (or gram). Despite these drawbacks, the plate-count method is a powerful means by which concentrations of viable organisms may be estimated. Also, if it is desirable to count a specific subgroup of microorganisms in a sample, selective media or special incubation conditions can often be used to encourage the growth of only this class of organism. More information is available in the chapter on quantitative microbiology.

As microbial quantitation involves the use of pipettes (or micropipettes) in preparing dilutions and inoculating plates, the beginning student in microbiology must become familiar with their use.

Due to the possibility of ingesting pathogens and toxic liquids, mouth-pipetting is forbidden in the laboratory! Pipettes are filled and subsequently emptied by the use of propipettes or other pipette bulb. Pay close attention to the demonstration of their use.

Important Safety Consideration: When fitting the pipette and pipette bulb together, use very gentle pressure!! Do not jam these items together! (Force is usually not the answer, a good general rule to live by in this lab and in life for that matter.) The glass pipette will probably break and possibly cause severe injury. Handle the pipette only at the top inch or so.

For volumes of 5 ml of less, micropipettes are often the tool of choice. Instruments are available that are capable of dispensing 5 ml all the way to less than 1 µl (one one-millionth of a liter). Micropipettes have made it possible to miniaturize many experiments and greatly decrease the cost of running them. They are also easy to use and can dispense volumes quickly, increasing the number of experiments that can be performed in a set amount of time. Micropipettes are the tool of choice for small volumes.

24

DETERMINATION OF MICROBIAL GROWTH

AIM OF THE EXPERIMENT

Spectrophotometric/Colorimetric determination of bacterial

REQUIREMENTS

- Nutrient broth
- Culture tubes (2)
- Cotton Plugs (2)
- Conical Flask (100ml) (1)
- Pipette (5ml) (1)
- Spirit Lamp /Burner

THEORY

Lactobacilli being inoculated into culture tube shows growth. During growth, bacteria multiply. Each bacterial cell behaves as a particle and with increasing number of cell due to growth; these cells obstruct passage of light transmittance. As a result the inoculated culture tube attains turbidity, while the uninoculated tube remains relatively transparent. Thus the O.D. value of the inoculated tube is an indicator of bacterial growth.

SAMPLE

Lactobacilli capsules are available in medicine shops, which contain lypholized bacterial cells.

PROCEDURE

1. Take 10 – 15 ml of nutrient broth in 2 clean culture tubes and plug them.
2. Sterilize the tubes in pressure cooker / autoclave for 15min.
3. Allow the tubes to cool.
4. Take a *Lactobacilli* capsule, open it and add content to a conical flask.
5. Add 20ml. of distilled water to a conical flask and disperse the lactobacilli thoroughly.
6. Take 2ml of the content of the conical flask by sterilized pipette and transfer to one of the plugged culture tube aseptically in presence of flame.

7. Let the other culture tube remain as uninoculated.
8. Incubate the tubes at 37°C or room temperature for 24hrs.

OBSERVATION

Uninoculated tubes will remain transparent, while the inoculated tube will become turbid due to bacterial growth.

ESTIMATION

Measure the optical density of the content of the inoculated tube at 660 nm by spectrophotometer / colorimeter, taking the content of the uninoculated tube as blank.

NOTE

- Spectrophotometric /colorimetric principles must be explained to students.
- Growing bacteria at different environmental temperatures can also perform these experiments and growth can be assessed in terms of O.D.

25

ISOLATION OF STAPHYLOCOCCI

AIM

Isolation of *Staphylococci* from skin surface

THEORY

Mannitol salt agar is selective medium containing high concentration of salt and Mannitol. Skin surface *Staphylococci* can tolerate high salt concentration and can ferment mannitol-producing acid. Hence these bacteria are favoured in the petriplate. Acid secreted by these bacteria change the pH (from 7.4 to 7.0) of the medium as a result phenol red changes to yellow colour.

REQUIREMENTS

- Petriplate (2)
- Cotton Swab
- Ringer Solution (20ml)
- Forceps
- Mannitol
- Beef Extract
- Peptone
- NaCl
- Phenol Red
- Distilled Water
- Agar
- Ringer Solution / Methanol
- Autoclave
- BOD Incubator
- Culture Hood

PREPARATION OF MANNITOL SALT AGAR MEDIUM

Dissolve the constituents [(Beef extract (10g), Peptone (10g), NaCl (75g), Mannitol (10g), Phenol Red (25g), Distilled Water (500ml)] and add to 15g agar dissolved in 500ml of distilled water and autoclave at 15lb pressure for 15min. Adjust the pH pf the medium to 7.4

PROCEDURE

1. Sterilize all the articles under UV lamp inside the culture hood.
2. Take the sterilized cotton swab and put it aseptically in 10ml Ringer Solution or Methanol.
3. Rub the cotton swab (by the help of forceps) on skin surface (elbow, nose, side, arm pit).
4. Take 15ml of molten medium in sterilized petriplate and allow the medium to tool, sterilize under UV light inside hood.
5. Rub the surface of the medium of petriplate with the cotton swab containing the bacteria inside the hood
6. Incubate the petriplates at 37°C for 2 – 3 days.

OBSERVATION:

The bacteria colonies of *Staphylococci* develop a yellow zone around them.

26

GRAM STAINING TECHNIQUE

AIM

Study of the response of bacteria to Gram stain

THEORY

Gram stain is a differential stain developed by the Danish physician, Hans Christian Gram. According to the response of the stain, eubacteria can be classified as Gram (+) or Gram (-). This differential response is due to the difference in the composition of their cell wall. In Gram (-) bacteria the cell wall has high lipid content. Thus the stain along with the lipid comes out when dissolved in alcohol. This makes the bacterial wall colourless and treated with counter stain (Safranin) bacteria becomes red coloured. While in Gram (+) bacteria, lipid content being negligible, alcohol treatment does not make any difference and the bacteria retain the primary stain complex and there is no scope of accepting the counter stain. These bacteria thus remain violet and concluded to be positive towards Gram stain. This technique is a basic technique for classifying bacteria and used in pathological diagnosis.

REQUIREMENTS

- Sterilized slides
- 95% ethyl alcohol
- Aqueous solution (5%) of crystal violet
- Lugol's Iodine (3g Iodine and 6g potassium Iodide in 900ml of distilled water)
- Acetone alcohol (1:1)
- Safranin (1%aqueous solution)
- Wash bottle
- Wire loop
- Burner
- Distilled water
- Petridishes
- Forceps
- Dropper/pipette
- Blotting paper
- Bacterial culture
- Microscope

BACTERIA SAMPLE

- Fresh curd water can be smeared to have *Lactobacilli,* which are Gram (+)
- Crushing root nodules of leguminous plant can be smeared to yield Gram (-) bacteria.

PREPARATION OF STAINS

1. Crystal Violet Stain

Solution A

- Crystal Violet   -   2g
- Ethyl Alcohol   -   20ml

Solution B

- Ammonium Oxalate   -   0.8g
- Distilled Water   -   80ml

Crystal Violet Stain is formed when solution A and B are mixed thoroughly.

2. Safranin

- Safranin (2.5% solution in 95% Ethyl Alcohol)   -   10ml
- Distilled Water   -   100ml

3. Lugol's Iodine

- Iodine   -   1g
- Potassium Iodide   -   2g
- Distilled Water   -   300ml

PROCEDURE (PROTOCOL BASED ON HUMPHRIES AND MURRAY, 1974)

1. Clean the slide with alcohol to make them grease free.
2. Pour a few loops full of distilled water in the center of the slide.
3. Touch the wire loop with the bacterial colony and transfer the bacteria to the glass slide, mix and spread evenly over the slide.
4. Hold the slide with a forcep and dry it by quickly passing it over the burner a couple of times, this will fix the bacteria to the slide.
5. Cover the smear with crystal violet stain for 30 –60 seconds.
6. Drain off the stain and add Lugol's Iodine solution over the smear for another 30 seconds. This will fix the stain onto the cells.
7. Wash the slide with ethyl alcohol until no more colour flows from the smear.
8. Wash the smear immediately with distilled water to prevent excessive decolourization.
9. Counter stain the smear with Safranin / Neutral Red for 30 – 60 seconds.
10. Let the stained slide air-dry.

OBSERVATION

Examine the slide under microscope with oil immersion lens. Note the colour of the bacterial cells.
- Gram (+) bacteria (e.g. *Lactobacillus* sp.) will appear violet or black.
- Gram (-) bacteria (e.g. *E. coli*) will appear red.
- Sketch the shape of the bacteria.

27

DETRMINATION OF MICROBIAL QUALITY OF MILK

AIM

Determination of the quality of milk by Methylene Blue Reductase test (MBRT)

THEORY

Milk is a heterogeneous mixture of carbohydrate, protein, fats, minerals and vitamins. It is usually contaminated by microbes, which when consumed can spread diseases like diphtheria, typhoid, etc. It is thus important to know the microbiological quality of this important food item before it can be safely consumed.Bacteria present in the milk utilize the oxygen and produce reduced environment. Methylene Blue (MB) is a colour sensitive chemical, which is blue when oxidized state and colourless when reduced. The speed of colour disappearance is proportional to the number of bacteria present and is taken as an indicator of bacterial load.

REQUIREMENTS

- Test tubes with cotton plug   –   3
- Sterile 10ml pipette   –   3
- Sterile 1ml pipette   –   3
- Distilled Water

SAMPLE

- Methylene Blue Solution (1:25,000)   –   3ml
- Three milk samples (from different sources)   –   100ml each

PROCEDURE

1. Prepare MB solution by dissolving 1mg of MB aseptically in 25 ml of distilled water.
2. Take 10ml of raw milk sample in a test tube using sterilized 10 ml pipette.
3. Add 1ml of MB solution to the milk sample and mix the content thoroughly.
4. Incubate the test tube in water bath or at 37°C for 6 hrs.

5. Observe the incubated milk tubes every 30 min. for the change in colour from blue to white. Record the time required for the decolourization.

OBSERVATION

Milk Sample	Time of Decolourization
A.	20min.
B.	160min.
C.	350min.

MBRT (Time)	Classification of Milk	Approx. number of Bacteria/ml of the sample
1. 1 to 30 min.	Very poor quality	$> 2 \times 10^7$
2. 31 to 120min	Poor quality	$> 4 \times 10^6$
3. 121 to 360 min	Fair Quality	$> 5 \times 10^2$
4. > 360 min	Good Quality	$< 5 \times 10^2$

ANIMAL CELL CULTURE

28

CELL/TISSUE CULTURE : BASIC CONCEPTS

BASIC REQUIREMENTS

Cultured cells are required in a spectrum of disciplines ranging from biochemistry to molecular and cell biology. Cells and tissue culture procedures are simple but require extreme care to avoid contamination.

MATERIALS NEEDED

- Dulbecco's modified Eagles Medium.
- Penicillin / Streptomycin.
- Sodium Bicarbonate ($NaHCO_3$).
- Fetal Calf Serum.
- Trypsin.
- Hoechst 33258 Glycerol.
- 0.47 Trypan Blue.
- $25cm^2$ and $75cm^2$ tissue culture flasks – 1ml, 10ml, 25ml.

EQUIPMENT REQUIRED

1. Laminar Flow Hood:

   - Vertical flow – sterile air blows down onto the work area
   - Room air is sucked in underneath work area
   - Laminar flow is lost when sash is too high
   - UV light used to help sterilize surfaces

2. Incubator:

   - Used to warm media

3. CO_2 Incubator:

   - Open dishes and multiwell plates
   - Maintain temp. and humidity
   - Control pH by CO_2 buffer (carbonic anhydrase)
   - Frequent cleaning essential

4. Sterilizer:

- Autoclave – sterilize non heat labile liquids
- Sterilize glassware and instruments

5. Sterilizing oven:

- Useful for glassware
- Avoid liquid contamination

6. Refrigerator and freezer:

- Good to store media and frozen serum

7. Inverted microscope:

- Needed to see morphological change
- Helps monitor early phases of contamination

8. Fluid handling:

- Pipettes- fixed volume and variable micropipettes
- Peristaltic pump

9. Consumables:

- Micropipette tips
- Regular Eppendorf
- Sterile plates and tubes
- Syringes and needles
- Culture vessels

10. Other stuff:

- Balance
- Haemocytometer
- PH meter
- Centrifuge
- Colorimeter
- Distilled water unit
- Vacuum pump

PROCEDURE FOR WASHING GLASSWARE

1. Regular Glassware washing

Prerinse the glassware with demineralised water use and immerse for 2 to 3 hrs. in a 1% solution of a mild detergent such as 7XPF. Scrub using soft brushes. 7XPF can be reused several times. After detergent treatment, the glassware is washed once in warm tap water, twice in DW and once in double DW. Thereafter, the glassware is dried in an oven at 160°C for 2hrs. Clean glassware should wet uniformly when exposed to a thin stream of DW.

2. Glassware that has come in contact with biological material

Add 5% solution of 7X detergent and autoclave. Thereafter follow steps as described above.

3. Glass Cover slips

Place the cover slips in a solution containing 60ml of alcohol and 40ml of HCl. Leave for at least 30 min. at room temperature. Afterward, rinse several times with DW. Dry each cover slip with tissue paper, place in a 100mm tissue culture dish and sterilize in an oven.

STERILIZATION

Solutions, glassware and other items can be sterilized by moist heat in an autoclave. Sterilize for 30 min. at a pressure of 15lb/in^2. Empty bottles should have cap loosely screwed, whereas partially filled bottles should have their caps tightly screwed on. The caps should be covered with a piece of aluminum foil and strip of indicator tape placed. When sterilization is complete, the pressure should be lowered gradually to avoid boiling of the liquid.

Pasteur Pipettes placed in glass cylinders covered with aluminum foil, other glassware as well as small items wrapped in aluminum foil can be sterilized by dry heat in an oven. Sterilization is for 90min. at 160°C. Indicator tape is used to assess sterility.

Plastic ware (open dishes) can be sterilized by exposure to an ultraviolet light (UV) germicide lamp.

DISPOSAL METHODS

1. Media

Aspirate into disinfectant trap and autoclave when filled. Autoclave for 60 min. at 15lb/in^2

2. Plastic ware

Place all disposable plastics (flasks, dishes) in a stainless – steel container and autoclave.

MAINTENANCE OF ASEPTIC ENVIRONMENT

Work area free from drafts- even working in a hood can introduce contaminated air

- People walking by
- Open windows
- Some lab equipments

Work area

- Clear surface and clean with methanol or 70%EtOH
- Keep area free from clutter (stuff you don't need immediately)
- Clean spills quickly
- Remove anything after use and swab area again

Work habits

- Wash hands
- Tie back hair
- Don't talk or work when sick

Sterile handling

- Swab everything –work area, bottles, flasks, etc.
- Use caps in bottles rather than stoppers
- Flame bottles when opened and pipettes
- Hold bottles and flask at right angle when open (straight up let things fall in)
- Use pipette aids properly (no mouth pipetting)
- Resist pouring from one container into another
- Clean all equipments frequently: hoods, incubator, even refrigerator

SERUM

Serum:

Extremely complex mixture

Contains

- food substances
- metabolites
- gases
- hormones
- plasma protein
- growth factors

Note*

Always use universally accepted cell culture supplement. Quantity and Quality of components are affected by age, health and nutrition of the animals used. Thus it's subject to biological variation. Best stuff obtained from fetal or newborn bovines.

Functions

Changes in the physiological properties of the medium:

- Viscosity
- Osmolality
- Diffusion rates
- Add protease inhibitors

Provides nutrients for metabolic processes

- Cholesterol
- Albumin
- Transferrin
- Appolipoproteins
- Attachment factors
- Growth factors
- Hormones

- helps solubilize essential nutrients that are normally insoluble
- contains detoxification proteins
- slow release of Fatty acids

Growth Factors

- Group of polypeptides with mitogenic activity
- Commercially synthesized (very expensive)
- Synergistic or additive activity
- Many have pharmaceutical importance, like:

    - EGF - Epidermal growth factor
    - FGF - Fibroblast growth factor
    - PDGF - Platelet derive growth factor
    - IGF - Insulin-like growth factor
    - Interferons

Hormones

- Growth hormone
- Insulin - improves plating efficiency
- Hydrocortisone - improves cloning efficiency

Attachment Factors

- Have affinity to fibrous proteins (collagen and fibrin)
- Enhance attachment of fibroblasts to collagen and synthetic surfaces
- Increase cell mobility
- Bind to variety of things:

    - Gelatin
    - Collagen
    - Fibrinogen
    - Heparin
    - Fibrin

Many cells attach to Poly-L-lysine coated surfaces

- Poly-L-lysine coats solid surfaces
- residual cationic sites bind to anionic sites on membranes

Fetal Bovine Serum (FBS)

- most widely used type
- most expensive
- has high levels of growth stimulatory factors
- has lower levels of growth inhibitory factors

Other Types

- Human Serum
- Bovine Calf Serum
- Newborn Bovine Serum
- Donor Bovine Serum
- Donor Horse Serum

Collecting and Testing of Serum

- Use cardiac puncture
- Umbilical exsanguination
- Get thorough clotting at refrigeration temp
- Separate by centrifugation

Selecting Serum Supplier

- one that processes it's own
- or obtained from veterinary approved collecting sites
- reputable ones provide documentation of source
- should welcome audit of facility

Country of Origin

- Important if cells are to be exported
- May have restrictions if serum from overseas
- Some countries livestock carry transmissible agents

Certificate of Analysis

- Lists QC tests on final product
- Batches should be free of mycoplasma and viruses

Performance Characteristics

- Cloning efficiency - promote growth of hybridoma cells lines
- Plating efficiency - use human transformed line to measure continuous cell lines
- Growth promotion - support proliferation of fibroblasts through multiple subculture

Serum Free Media

Used to help understand role of serum, nutrients, hormones and growth factors
Economy - serum is expensive
Can eliminate serum if:

- Optimize nutrient concentrations
- Balance ratio concentrations of nutrients

Nutrient mixtures require:

- Fat and water soluble hormones

- Growth factors
- Attachment factors
- Transports proteins

Make up of mixture dependant on:

- Cell type
- Transformed or non-transformed
- Basal medium used
- Cell density
- Physio-chemical environment
- Dissolved O2
- Aeration
- Temp
- pH
- Shear sensitivity

Adapting cells to Serum Free Media

Slowly lower serum concentration over time
Add serum substitutes
Add defined proteins
Use a rich medium that supports cell proliferation

Advantages

- Defined
- Consistent from one batch to the next
- Constant supply
- Can be optimized for growth of specific cell types
- Can be optimized to promote product secretion
- No need to worry about growth inhibitors found in serum
- Easy to purify

Disadvantages

- Expensive
- Short shelf life
- May need trypsin inhibitors
- May be cell type specific
- May require adaptation period
- Cells are more subject to damage

TYPES OF SERUM FREE MEDIA

1. Totally protein free

- Most desirable
- Hard to produce

2. Contain serum substitutes

- HL1
- CPSR1-5
- Nutridoma+
- Nuserum

3. Contain defined proteins and additives

Media Formulations:

Defined medium

The chemical structure and the concentration of every component is known.

Undefined medium

Contain one or more unidentified components.

Serum Free Media

Used to help understand role of serum, nutrients, hormones and growth factors
Economy - serum is expensive
Can eliminate serum if:

- Optimize nutrient concentrations
- Balance ratio concentrations of nutrients

Nutrient mixtures require:

- Fat and water soluble hormones
- Growth factors
- Attachment factors
- Transports proteins

Make up of mixture dependant on:

- Cell type
- Transformed or non-transformed
- Basal medium used
- Cell density
- Physio-chemical environment
- Dissolved O2
- Aeration
- Temp

- pH
- Shear sensitivity

Adapting cells to Serum Free Media

- Slowly lower serum concentration over time
- Add serum substitutes
- Add defined proteins
- Use a rich medium that supports cell proliferation

Advantages

- Defined
- Consistent from one batch to the next
- Constant supply
- Can be optimized for growth of specific cell types
- Can be optimized to promote product secretion
- No need to worry about growth inhibitors found in serum
- Easy to purify

Disadvantages

- Expensive
- Short shelf life
- May need trypsin inhibitors
- May be cell type specific
- May require adaptation period
- Cells are more subject to damage

TYPES OF SERUM FREE MEDIA

1. Totally protein free

- Most desirable
- Hard to produce

2. Contain serum substitutes

- HL1
- CPSR1-5
- Nutridoma+
- Nuserum

3. Contain defined proteins and additives

BASAL SALT MIXTURES (BSM)

Contain a mixture of:
- Calcium Chloride
- Magnesium Chloride
- Potassium Chloride
- Potassium Phosphate Mono basic (anhydrous)
- Sodium Chloride
- Sodium Phosphate (Dibasic)

Functions of BSM:

- Provide water and certain bulk ions for cell metabolism
- Maintain intra and extra cellular osmotic balance
- Provide a buffering system to maintain the medium within the physiological pH range

The Most Common Types of Media

Medium 199	Morgan *et.al.*, 1950
MEM	Eagle *et.al.*, 1957
CMRL 1066	Parker *et.al.*, 1957
DMEM	Dulbecco, 1959
Ham's F-12	Ham, 1965
RPMI 1640	Moore *et.al.*, 1967

BASIC COMPONENTS OF MEDIUM

Low Molecular Weight Nutrients:
- Nitrogen sources
- Vitamins
- Bulk Ions
- Lipids and Phospholipd precursors

Non-Nutrient Substances:

- Antibiotics
- Reducing agents
- Buffers
- Phenol Red
- Protective Agents
- Attachment Factors

WATER QUALITY

Demineralised, double distilled water should be used.

Types of contaminants in water:

- Inorganic: Heavy metals, iron, calcium, chlorine

- Organic: Detergents or by-products of plant decay
- Bacterial products: Endotoxins: Lipopolysaccharides generated by Gram Negative Bacteria

LOW MOLECULAR WEIGHT NUTRIENTS

Energy Sources: Carbohydrates – Glucose

Nitrogen Sources: Amino Acids

ESSENTIAL AMINO ACIDS

- Arginine
- Lysine
- Histidine
- Isoleucine
- Cystein
- Methionine
- Phenylalanine
- Threonine
- Tryptophan
- Tyrosine
- Valine
- Glutamine

Non-Essential Amino Acids:

- Alanine
- Asparagine
- Asparatic Acid
- Glutamic Acid
- Glycine
- Proline
- Serine·

IMPORTANCE OF AMINO ACIDS

- Differentiation may lead to reduction of cell capacity to synthesize amino acids
- Biochemical specialization lost *in vitro*
- Failure to synthesize a particular AA *in vitro* may represent the lack of correct precursor or cofactor in artificial environment.
  Ex. Monkey kidney cells have reduced ability to convert folic acid. The latter is required as a cofactor for the conversion of Serine to Glycine.
- Monolayer cells divide at a very active pace, thus insufficient amounts can be synthesized.

IMPORTANCE OF GLUTAMINE

- Support cell growth and AA uptake
- Support utilization

- A source of energy
- Support protein turnover

VITAMINS

A. Water Soluble

- Biotin
- Niacinamide
- Pyridoxine
- Thiamine
- Ascorbic Acid
- Folic Acid
- Riboflavin
- Vitamin- B12
- Pantothenate

B. Fat Soluble

- Vitamin A
- Vitamin D
- Vitamin E
- Vitamin K
- E and K are toxic when in exces

BULK IONS

Sodium: Maintains osmotic pressure in the medium
Potassium: Maintains osmotic pressure in the cell
Calcium & Magnesium

- Essential for intracellular enzymes
- Participate Cell attachment and spreading
- Calcium is essential for cytoplasm progress

Iron: Needed for respiratory pigments – cytochromes
Carbonate

- Natural buffering
- Participate in basic biochemical processes.

Phosphate: Energy carrier

FACTORS AFFECTING MEDIA

pH

- Most cells grow well at pH 7.4:

        Some fibroblasts grow at 7.4-7.7, transformed cells 7.0-7.4, epidermal cells sometimes at 5.5

- Phenol Red Indicator used:

    Purple at pH-7.8, Pink at pH-7.6, Red at pH-7.4, Orange at pH-7.0, Yellow at pH-6.5
- CO_2 and Bicarbonate Buffering system:

    Must use buffers to prevent swings in pH, related to equations catalysed by carbonic anhydrase

O_2 Levels

- Cells in culture rely mostly on Glycolysis
- Rely on dissolved oxygen
- Too much leads to the formation of free radicals

Use of Reducing Agents

B-Mercaptoethanol:

- Stimulate cysteine uptake by forming a mixed disulphide
- Help prevent peroxide damage by restoring the reduced form of glutathion

Temperature:

- Maintain close to 37°C
- Better lower than over
- Cells die at 40°C
- Tip: keep temperature constant by getting in and out of incubator quickly

CULTURING AND SUB-CULTURING OF ANIMAL CELLS

Subculture

Anchorage dependent cells exhibit strong to very light adherence depending on cell type:

- Dog Kidney Cell Line MDCK adheres very strongly and requires both a protease (Trypsin) and a chelating agent (EDTA) to detach
- CHO cells adhere very lightly and can be loosened by gentle tapping of the culture vessel.
- The majority of adhesion cells require at least the addition of a protease to dissociate the cells from the vessel.

Trypsin 0.25% + 1mM EDTA (pH 7.5) is normally used for dissociation of cells.

- Alternates to Trypsin:

    - Pronase
    - Dispase

Trypsin Treatment

The time required to dissociate cells from the culture vessel surface is dependent on:

- Cell Type
- Population density
- Serum concentration in the growth medium (contains protease inhibitors)

- Potency of the trypsin solution
- Time since last subculture

Quantitation of Cells

Important for routine maintenance:

- Best time for subculture
- Consistency
- Optimum dilution
- Estimated plating efficiency at different densities
- Testing medium
- Testing serum

GROWTH OF ANIMAL CELLS IN CULTURE

1. Anchorage-dependent

- *Require a surface to attach to and spread to grow and differentiate.*
- T-Flasks Roller Bottles
- Microcarriers
- Hollow fiber

2. Anchorage-independent (Suspension)

- Capable of growth and differentiation without attachment
- Spinner Flasks
- Stirred-Tank Bioreactors
- Air lift Bioreactors

FACTORS AFFECTING CELL DENSITY AND VIABILITY

- Concentration of limiting factors
- Concentration of growth factors
- Concentration of toxic metabolites
- Contact inhibition

CHANGES IN THESE FACTORS CAN CAUSE

- Decreased rate of enzyme and protein production
- Alterations in cellular metabolism
- Morphological changes
- Alterations in cellular proliferation

29

PREPARATION OF MEDIUM

METHOD

1. Sterilize laminar flow hood by UV irradiation for 45min.
2. Take all the materials inside the flow hood.
3. Take 500ml of sterile double distilled water in a 1000ml measuring cylinder
4. Transfer the contents of the powdered medium into a 1 litre-measuring cylinder and add $NaHCO_3$ (400mg) and Gentamycin (50mg).
5. Mix thoroughly to dissolve the powdered medium, $NaHCO_3$ and Gentamycin.
6. Fill the cylinder to 1 litre mark with D.D.water, mix and transfer to sterile 2-litre flask and mix. Pinkish red colour of the medium indicates normal pH range.
7. Assemble the filter sterilization set-up and carry out the filtration under negative pressure.
8. Prepare 400ml of medium containing 10% Adult Bovine Serum using 100ml measuring cylinder and store in a 500ml sera lab bottle.
9. Transfer the remaining medium without serum into big glass bottles.
10. Store the medium in refrigerator, dispose the used membrane and immerse the used glassware in water for washing.

Composition of medium

Amino Acids	mM	Vitamins	µM	Salts	mM	Other materials	
Arginine	0.1	Biotin	1.0	NaCI	100	Glucose	5mM
Cystein	0.05	Choline	1.0	KCI	5	Penicillin	0.005%
Glutamine	2.0	Folic Acid	1.0	NaH_2PO_2	1	Streptomycin	or
Histidine	0.05	Nicotinamide	1.0	$NaHCO_3$	20	Gentamycin	0.005%
Isoleucine	0.2	Pantothenic Acid	1.0	$CaCI_2$	1	Goat or calf serum 5%	
Leucine	0.2	Pyridoxal	1.0	$MgCI_2$	0.5		
Lysine	0.2	Thiamine	1.0				
Methionine	0.05	Riboflavin	0.1				
Phenylalanine	0.1						
Threonine	0.2						
Tryptophan	0.02						
Tyrosine	0.1						
Valine	0.2						

30

Commercially available fetal calf serum, newborn calf and horse serum are used in tissue culture medium. Following are some important functions of serum in supporting the growth and the proliferation of different varieties of cells:

1. Carrier / buffer / chelator for labile or water insoluble nutrients.
2. Binds and neutralizes toxins.
3. Provides protease inhibitors, which inactivates trypsin
4. Assists attachment of cells to substrate.
5. Provide essential low molecular weight nutrients.
6. Provides hormones and peptide growth factors.

SERUM

Extremely complex mixture
Contains:

- Food substances
- Metabolites
- Gases
- Hormones
- Plasma protein
- Growth factors

FUNCTIONS

Provides nutrients for metabolic processes

- Cholesterol
- Albumin
- Transferrin
- Apolipoproteins
- Attachment factors
- Growth factors
- Hormones
- Helps solubilize essential nutrients that are normally insoluble
- Contains detoxification proteins
- Slow release of Fatty acids

- Sometimes toxic in high levels
- Bind up and slowly release them into solution

GROWTH FACTORS

Many have pharmaceutical importance

- EGF - Epidermal growth factor
- FGF - Fibroblast growth factor
- PDGF - Platelet derive growth factor
- IGF - Insulin-like growth factor
- Interferons

HORMONES

- Growth hormone
- Insulin - improves plating efficiency
- Hydrocortisone - improves cloning efficiency

ATTACHMENT FACTORS

- Have affinity to fibrous proteins (collagen and fibrin)
- Enhance attachment of fibroblasts to collagen and synthetic surfaces
- Increase cell mobility
- Bind to variety of things: Gelatin, Collagen, Fibrinogen, Heparin, Fibrin, etc. Many cells attach to Poly-L-lysine coated surfaces
- Poly-L-lysine coats solid surfaces
- Residual cationic sites bind to anionic sites on membranes

FETAL BOVINE SERUM (FBS)

- Most widely used type
- Most expensive
- Has high levels of growth stimulatory factors
- Has lower levels of growth inhibitory factors

OTHER TYPES

- Human Serum
- Bovine Calf Serum
- Newborn Bovine Serum
- Donor Bovine Serum
- Donor Horse Serum

PREPARATION OF GOAT SERUM:

All steps should be carried out on ice unless otherwise stated.

1. Collect 1 litre of goat blood aseptically from slaughterhouse.

2. Keep at room temperature for 30 min. for clotting
3. Keep at 4°C for 2-3hrs. for the clot to sink
4. Transfer the clear serum to 500ml flask.
5. Centrifuge at 5000rpm for 10 min. at 4°C.
6. Inactivate the complement components by keeping the serum bottle in a water bath at 56°C for 30 min.
7. Allow it to come to room temperature and store it at – 20°C for further use.
8. Sterilize the serum by passing through Millipore membrane and transfer into sterile serum bottles.

31

PRIMARY CELL CULTURE

THEORY

Culture started fresh from tissues is called primary culture. Almost any tissue can be cultured, though high rates of success in cell culture is often recorded in case of embryonic and tumour tissues rather than in normal adult tissues.

SUITABILITY OF MATERIALS FOR DOING PRIMARY CULTURE

- Types of issue
- Species of origin
- Age of the animal
- Dissociation medium used
- Enzymes used
- Enzyme concentration and purity
- Temperature
- Incubation time

SEQUENCE

1. Isolation of tissue
2. Dissection and /or disaggregating

    - Cells migrate out from small tissue chunks in flask
    - Disaggregating either mechanically or enzymatically
    - Create cell suspension
    - Some cells will adhere to substrate in culture

3. Culture in flasks

DISSECTING AND /OR DISAGGREGATING CELLS /TISSUE IS ACCOMPLISHED IN ONE OF THREE WAYS

1. Explants:

    - Original method developed at turn of the century
    - Finely chop tissue to very small parts

- Cells migrate from pieces onto culture flask
- Useful with small tissue samples (biopsies)

Slow process, certain cells migrate faster than others (fibroblasts)

2. Mechanical Disaggregation:

- Vigorous pipetting
- Pressing tissue into a mesh
- Wash cells through sieve
- Faster than enzymes but less yield
- Causes mechanical damage to cells

3. Disaggregating by Enzymatic action:

- Need to disrupt cell-cell adhesion proteins
- Crude preparations of enzymes are often used
- May contain other proteases as contaminants – work better
- Purified mixtures are less damaging to cells
- Cells can be damaged to point of lysis

ENZYMES FREQUENTLY USED FOR DISAGGREGATING CELLS

- Trypsin (Strongest)
- Papain
- Elastase
- Hyaluronidase
- Collagenase Type I
- Collagenase Type II
- Collagenase Type III
- Collagenase Type IV
- DNAse (Weakest)

COMMON FEATURES OF TISSUE ISOLATION

- Fat or necrotic tissue should be removed
- To minimize damage chop tissue finely with sharp instrument
- Remove disaggregation enzymes by centrifugation
- Higher concentration of cell needed for primary isolation than sub culturing
- Use richest medium possible
- Embryonic tissue is best

  - Disaggregates better
  - Get more viable cells
    Proliferates better

ISOLATION OF EPITHELIAL TISSUE OR CONNECTIVE TISSUE

1. Sources of Epithelial Tissue:

- Epidermis

- Glands of the skin (sweat or sebaceous)
- Outer corneal layer
- Lining of digestive and reproductive tracts
- Endothelium
- Parts of liver, pancreas, pituitary, gastric and intestinal glands

2. Sources of Connective Tissue:

- From the dermis of the skin
- Capsule and stroma of various organs
- Mucous and serous membranes
- Cartilage, bone, tendons, ligaments, adipose

Components of Connective Tissue

Cells	Fibres
• Fibroblasts	• Collagen
• Adipocytes	• Reticular fibres
• Neutrophils	• Elastic fibres
• Lymphocytes	
• Monocytes	
• Macrophages	

YIELD AND VIABILITY ISSUES

A. Low yield and Low Viability

- Under or over dissociation
- Have cellular damage

Solution

- Use less harmful digestive enzyme
- Prefer Trypsin to Collagenase
- Decrease concentration of enzyme

B. Low yield and high Viability

- Under dissociation

Solution

- First, increase concentration of enzyme, or
- Increase incubation time
- Next, use other enzyme or combinations

C. High Yield and Low Viability

- Good dissociation
- Have cellular damage 9enzyme too strong)

Solution

- Reduce enzyme concentration
- Reduce incubation time
- Dilute enzyme with BSA or soy protein

D. High Yield and High Viability:

- Good dissociation
- Good yield

PRIMARY CULTURE OF MOUSE CELLS

Protocol

1. Sacrifice the mouse by cervical dislocation. Place the animal in supine position.
2. Swab the abdomen with 70% ethanol and cut open along the mid-ventral line.
3. Remove spleen / liver and transfer into a beaker containing PBS. Transfer into flow hood immediately.
4. Transfer the live / spleen into petriplate containing PBS inside the flow hood.
5. Wash tissue with PBS.
6. Transfer the remains to another petriplate containing small amount of PBS and mince thoroughly with a pair of bent scissors.
7. Transfer the minced tissue to the trypsinization flask containing 40ml of 0.25% trypsin in PBS.
8. Stir the contents at 37°C for 30 – 60 mins.
9. At the end of the period add 5ml of medium containing serum and stir the contents fir 2 mins. To inactivate the action of trypsin.
10. Filter the cell suspension through sterile cheesecloth and collect the filtrate into a 100 ml conical flask.
11. Centrifuge the filtrate at 1000 rpm for 10 mins.
12. Pour out the supernatants and resuspend the pellet in 5ml of medium. Distribute   equally to all the 120cm2 culture bottles and incubate at 37°C.
13. Live cells will attach to the surface of the culture bottle in about 3 – 4 hrs.
14. Next day discard the medium containing floating dead cells and wash the cells with PBS. Replace with 5ml of fresh medium containing 10% serum. Replace the medium every third day and observe the cells under the microscope each day.
15. Sub-culture the cells once they reach confluency (in about a week).

32

CULTURING AND SUB-CULTURING OF ANIMAL CELLS

SUBCULTURE

Primary culture of mouse embryo cells consists of a diverse population of cell types. Subsequent passaging of the cells results in the selection of distinct homogenous populations of cells. The cell population that develops on serial passaging of the culture is dependent on the medium used for culturing. Careful selection of specific medium may allow for the selection of cell populations of interest. Cultures derived from the sub-culture of primary culture are called 'secondary cultures'.

PROCEDURE FOR SUB-CULTIVATION

- Take one TC bottle containing a fully formed Monolayer of cells
- Discard the old medium and wash the Monolayer thrice with PBS.
- Add 5 – 6 drops of 0.25% Trypsin and allow the drops to spread over the entire Monolayer.
- Wait for a minute and shake the bottle vigorously to facilitate the cells to come off from the substratum.
- Once the cells start coming off add 5ml of medium containing serum.
- Flush with Pasteur pipette to dislodge the cells adhering to the glass surface.
- Divide the cell suspension into two bottles incubate at 37°C.

Anchorage dependent cells exhibit strong to very light adherence depending on cell type:

- Dog Kidney Cell Line MDCK adheres very strongly and requires both a protease (Trypsin) and a chelating agent (EDTA) to detach
- CHO cells adhere very lightly and can be loosened by gentle tapping of the culture vessel.
- The majority of adhesion cells require at least the addition of a protease to dissociate the cells from the vessel.

Trypsin 0.25% + 1mM EDTA (pH 7.5) is normally used for dissociation of cells:

- Alternates to Trypsin:
- Pronase
- Dispase

TRYPSIN TREATMENT

The time required to dissociate cells from the culture vessel surface is dependent on:

- Cell Type
- Population density
- Serum concentration in the growth medium (contains protease inhibitors)
- Potency of the trypsin solution
- Time since last subculture

QUANTITATION OF CELLS

Important for routine maintenance:

- Best time for subculture
- Consistency
- Optimum dilution
- Estimated plating efficiency at different densities
- Testing medium
- Testing serum

GROWTH OF ANIMAL CELLS IN CULTURE

1. Anchorage-dependent:

   - *Require a surface to attach to and spread to grow and differentiate.*
   - T-Flasks Roller Bottles
   - Micro carriers
   - Hollow fiber

2. Anchorage-independent (Suspension) :

   - Capable of growth and differentiation without attachment
   - Spinner Flasks
   - Stirred-Tank Bioreactors
   - Air lift Bioreactors

FACTORS AFFECTING CELL DENSITY AND VIABILITY

- Concentration of limiting factors
- Concentration of growth factors
- Concentration of toxic metabolites
- Contact inhibition

CHANGES IN THESE FACTORS CAN CAUSE

- Decreased rate of enzyme and protein production
- Alterations in cellular metabolism

- Morphological changes
- Alterations in cellular proliferation

PRESERVATION OF CELLS AND THEIR REVIVAL

Preservation cannot be done at Passage 1 (P-1) but from P-4 to P-100 or more. Cells are preserved in 30% glycerol or Dimethyl Sulphoxide (DMSO). DMSO is toxic, hence should be handled with extreme care.

PROCEDURE

1. Take 300 µl of cells.
2. Mix 200µl of 30% glycerol, so that the final concentration is 40%
3. Mix and seal the flask with paraffin.
4. Allow it slow freezing at 4°C overnight.
5. Transfer it to freeze at -20°C overnight.
6. Permanently store it at -70°C

REVIVAL PROCEDURE

For revival immediately transfer to 37°C by rapid thawing. Add 4ml of MEM in a culture flask. For DMSO wash twice and each time centrifuge at 2000 rpm. This will be P-21 after revival.

33

VITAL STAINING AND VIABLE COUNTING

AIM

Estimation of Viability by Dye uptake

PRINCIPLE

Viability assays are used to measure the proportion of viable cells following disaggregation, cell separation or freezing or thawing. Viability test of cells are dependent upon breakdown of membrane structure so that become impermeable to certain dyes like, naphthalene bleck, trypan blue, and a number of other dyes.

OUTLINE

Mix a cell suspension with stain, and examine it by low-power microscope

MATERIAL REQUIRED

Sterile or aseptically prepapred

- Cultured Cells
- Growth medium
- 0.25% trypsin
- PBSA

Nonsterile

- Haemocytometer
- Viable stain (e.g., 0.4% trypan blue or 1% naphthalene black in PBSA)
- Pasteur pipettes
- Microscope
- Tally Counter (if possible)

PROTOCOL

1. Prepare a cell suspension at a high concentration ($\sim 10^6$ cells/ml) by trypsinization or by centrifugation and resuspension.

2. Take a clean haemocytometer slide and place the coverslip
3. Mix one drop of cell suspension and one drop of trypan blue or four drops of naphthalene black stain.
4. Load the mixture in the counting chamber of the haemocytometer slide
5. Leave the slide for 2-3 min.
6. Place the slide under a microscope.
7. Count the the total number of cells and the number of stained (dead) cells, using a x10 objective.

ANALYSIS

Calculate the percentage of unstained cells which will give the percentage viability.

34

CHROMOSOME KARYOTYPING

AIM

Chromosome preparation from cultures tissue cells

PREPARATION OF CELLS

1. Set up the primary culture from spleen / liver cells as previously described.
2. At passage 4, when the cells sub-confluent add colchicines solution to a final concentration of 0.1mg/ml.
3. Incubate for 2-4hrs.
4. Remove the medium and wash the cells twice with warm PBS.
5. Remove the cells from the flask with 0.25% Trypsin and inactivate the Trypsin by adding 10% serum containing medium
6. Collect the cells in a 10ml test tube.
7. Centrifuge at 1200 rpm for 4 min and discard the supernatant.
8. Resuspend the cell pellet in 5ml of warm hypotonic (0.075M) KCl and incubate at 37°C for 15 mins.
9. Centrifuge as above and resuspend the cells in 3:1 methanol; glacial acetic acid mixture (fixative). Add fixative drop by drop slowly, while gently shaking the tube.
10. Leave at room temperature for 10 mins.
11. Centrifuge as above, discard fixative and add fresh fixative (now you can add fast).
12. Repeat the above step twice.
13. Resuspend the pellet in 1ml of fresh fixative.
14. Prepare the slides by air-drying procedure.

STAINING

Chromosomes can be stained with 1% Acetic Orcein, Acetocarmine, Carbol Fuchsin, and Giemsa.

GIEMSA STAINING PROCEDURE

1. Immerse the slides in Giemsa staining solution for 1.5 to 2 mins.
2. Wash the slides in alcohol grades (90 – 100%).
3. Dry the slides and observe under microscope (immersion oil); count the number of chromosome and take photograph.

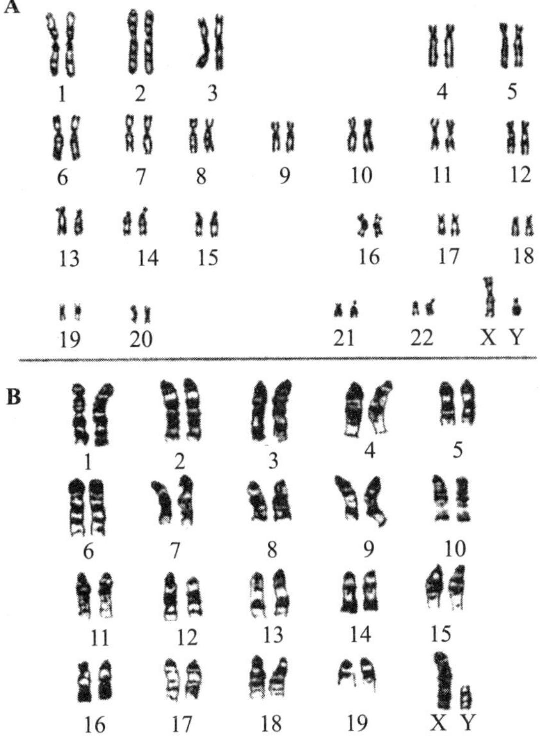

Fig. 22 Chromosome-Karyotypes

BIOCHEMICAL TECHNIQUES

35

pH AND BUFFERS

PRINCIPLE

Acids and bases

The word acid comes from the Latin word acere, which means 'sour'. All acids taste sour. The word alkaline could be from the Arabic al-qily, which means, 'to roast in a pan" or "the calcinated ashes of plants'. The word "base" might come from a Latin word, bassus, which means low. There are numerous definitions of acids and bases, most of which relate with solutions containing water.

- An acid is a proton donor or electron acceptor.
- A base can remove a proton from an acid or proton acceptor or electron donor.

STRONG ACIDS AND BASES

Strong means a substance that can be 100% ionised. This happens when water molecules are much stronger than the base in the acid. (Strength is not concentration, which is often expressed for acids as the number of moles per unit decimetre). All the acid dissociates, and all its protons are in solution, hence at the end no acid molecules are left. Examples of strong acids include;

$$\text{Hydrochloric acid (HCl} + H_2O \rightarrow H_3O^+ + Cl^-$$

$$\text{Sulphuric acid (H}_2SO_4 + 2\,H_2O \rightarrow 2\,H_3O^+ + SO_4^{-2}$$

$$\text{Nitric acids (HNO}_3 + H_2O \rightarrow H_3O^+ + NO_3^-$$

Identically, there will be no un-dissociated base molecules in solutions of strong bases. One should remember that K is always high in comparison to concentrations, and in the case of strong acids and bases in dilute solutions do not involve equilibrium,.

WEAK ACIDS AND BASES

Sometimes a few of the acids are weak and undergo partially dissociation to produce H_3O^+ and their bases or anions (A$^-$). The reason is that water cannot bind all the protons generated fron dissociation of such acids. As a result some of the hydrogen still remains undissociated and thus a measurement of $[H^+]$ does not give the total hydrogen ion concentration. However, if the acid is titrated against a dilute base, the undissociated molecules progressively dissociate and eventually the total hydrogen ion concentration can be found. Hence the "titratable acid" is different from the actual acidity (pH). With strong acids, the titratable acid and acidity are the same.

We talk of equilibrium in case of weak acids and bases, because the reactants are not finished. At equilibrium, the number of products being formed is exactly equal to the number being broken down to re-form reactants. That is, the rate of the forward reaction is exactly equal to the rate of the reverse reaction. Therefore, the arrows must be equal to indicate that the forward rate equals the reverse rate. For example, a weak acid (HA) dissociates reversibly in water:

$$HA + H_2O \longleftrightarrow H_3O^+ + A^-.$$

If we rearranging the equation to isolate the hydrogen ion concentration:

$$[H_3O^+] = K\,[HA]\,[H2O]/\,[A] \longrightarrow [H] = K\,[HA]/\,[A]$$

Assuming that $[H_2O]$ is in excess and therefore constant; and that **$[OH\text{-}]\,[H_3O^+] = 10\text{-}14$**. Therefore, **log** $\{H^+\} = \log K + \log \{HA\} - \log [A^-]$, and $-\log [H^+] = \log [A^-] - \log K - \log \{HA\}$

Acid-base reactions tend towards weaker acids and bases
For example:

$$\mathbf{HCl + H_2O \longleftrightarrow H_3O^+ + Cl^-.}$$

In general, in the above acid-base reactions, equilibrium favours the production of the weaker acid (e.g. H_3O^+) and base (e.g. Cl$^-$). Put simply, the acid produced has a stronger base than that in the acid, which lost the proton (i.e. H_2O base is stronger than Cl-). We may in fact consider acid-base reactions as a competition in which bases fight for protons.

WHAT IS pH?

The concept of pH was introduced in 1909 by the Danish chemist Søren Sørensen as a convenient way of expressing acidity. The reasons for the introduction of pH include:

- The slow colour-change tests, and a need for electrical methods of determining acidity or basicity;
- It is easier to write and compare pH than actual concentrations; e.g., "the pH varies from 2 to 13", instead of $[H_3O^+]$ varies from about 0.01 M to 0.0000000000001 M. Please note that comparisons of pH (or even of acidity) are limited to the use of the same conditions.
- The pH scale based on Lowry- Brønsted model is minimised to 0 to 14 because it assumes the presence of excess water.

The letters pH are an abbreviate for "pondus hydrogenii" (translated as potential hydrogen) meaning hydrogen power as acidity is caused by a predominance of hydrogen ions (H^+). Sørensen used to write pH as PH. The Compact Oxford English Dictionary states that the modern notation 'pH' was first adopted in 1920 by W. M. Clark (inventor of the Clark oxygen electrode) for typographical convenience, and claims that the "p" stands for the German word for "power", potenz, so pH is an abbreviation for "power of hydrogen". "p-Functions" have also been adopted for other concentrations. For example, "pCa = 5.0" means a concentration of calcium ions equal to 10-5 M.

From $-\log \{H+\} = \log [A-] - \log K - \log [HA]$ above, substitute $-\log[H]$ for pH, and $-\log K$ for pK. pH is the familiar measure of acidity or alkalinity, and pK is a measure of dissociation in water (remember we removed $[H_2O]$ and [OH-] from that equation - they are part of pK). By definition, pH = $-\log$ $[H_3O^+]$(or in this case $-\log$ [H+]. So, the definition of pH inherently assumes the presence of water in excess, and the factor 10-14 within pK.

A definition that uses the hydrogen ion activity:

In other cases, pH has been defined as pH = $-\log$ aH+ where aH+ is the hydrogen ion activity. In solutions that contain other ions and under varied conditions, activity and concentration are not the same. The activity indicates the hydrogen ions that are active, rather than the true concentration; it accounts for the fact that other ions and conditions surrounding the hydrogen ions might shield them and affect their ability to participate in chemical reactions.

THE BASIC PRINCIPLE OF ELECTROMETRIC pH MEASUREMENT

It is the determination of the activity of the hydrogen ions by potentiometric measurement using a standard hydrogen electrode and a reference electrode. The hydrogen electrode consists of a platinum electrode across which hydrogen gas is bubbled at a pressure of 101 kPa. Because of difficulty in its use and the potential for poisoning the hydrogen electrode, the glass electrode commonly is used. The electromotive force (emf) produced in the glass electrode system varies linearly with pH. This linear relationship is described by plotting the measured emf against the pH of different buffers. Sample pH is determined by extrapolation.

Because single ion activities such as a_H^+ cannot be measured, pH is defined operationally on a potentiometric scale. The pH measuring instrument is calibrated potentiometrically with an indicating (glass) electrode and a reference electrode using buffers having assigned values so that:

$$pH_B = - \log_{10} aH^+$$

where:

$$pH_B = \text{assigned pH of buffer.}$$

The operational pH scale is used to measure sample pH and is defined as:

$$pH_x = pH_B \pm \frac{F(E_X - E_X)}{2.303 \, RT}$$

where:

$$pH_x = \text{potentiometrically measured sample pH,}$$

$$F = \text{Faraday: } 9.649 \times 104 \text{ coulomb/mole,}$$

$$E_x = \text{sample emf, V,}$$

$$E_s = \text{buffer emf, V,}$$

$$R = \text{gas constant; } 8.314 \text{ joule/(mole }°K), \text{ and}$$

$$T = \text{absolute temperature, }°K$$

The symbol pH is derived from the French for "hydrogen power." The numbers in the pH scale stand for negative logarithms (10^{-n}) of the concentration of hydronium ions (H_3O^+), here verbally simplified to hydrogen ions (H^+) expressed in moles per liter of water. To obtain a mole of any substance, you simply measure out its molecular weight in grams. One gram of hydrogen atoms is a mole; 44 grams of CO_2 is a mole; the atomic weight of C = 12, that of O = 16, and 12 + 2(16) = 44. A mole of any substance contains the same number of particles — Avogadro's number, 6.02×10^{23}. The liter is a particular volume of water.

Thus a pH of 0 means a hydrogen ion concentration of 10^0 molar or moles per liter, which is one mole.

pH   Concentration of H$^+$ ions in moles per liter

pH				
0	10^0	=	1.0	molar
1	10^{-1}	=	0.1	molar
5	10^{-5}	=	0.00001	molar
7	10^{-7}	=	0.0000001	molar
10	10^{-10}	=	0.0000000001	molar

Since normal dissociation of water molecules into H$^+$ and OH$^-$ yields 10^{-7} moles of H$^+$, pure water has a **pH of 7**, the **neutral** point in the scale. Since the pH scale is logarithmic, a difference in 1 unit is a 10-fold increase or decrease in concentration of hydrogen ions.

I. MEASURING pH

Your instructor will demonstrate the use of the pH meter and of litmus paper.

Dilute each solution by half (add an equal volume of water). Measure and record the pH of each of the following aqueous solutions. Be sure to rinse the electrode with distilled water between each measurement and dry it carefully; use a twist of paper towel like dental floss to clean the groove in its tip.

Put a drop of each solution on pink and on blue litmus paper and record the color change. What is the effect of an acid and of a base on each color?

The pH of a buffered solution is calculated by using the Henderson-Hasselbalch equation:

$$pH = -\log K_a + \log {}^{[base]}/_{[acid]}$$

where K$_a$ is the acid dissociation constant of the weak acid in the buffered solution.

Example: What's the pH of a solution that contains 0.100 M acetic acid and 0.200 M sodium acetate? K_a ($C_2H_3O_2H$) $= 1.75 \times 10^{-5}$.

Solution: Using the Henderson-Hasselbalch equation, we find that:

$$pH = -\log K_a + \log {}^{[base]}/_{[acid]}$$

$$pH = -\log(1.75 \times 10^{-5}) + \log {}^{[0.200]}/_{[0.100]}$$

$$pH = 4.76 + 0.301$$

$$pH = 5.06$$

The Henderson-Hasselbalch equation in chemistry describes the derivation of pH as a measure of acidity (using pK$_a$, the acid dissociation constant) in biological and chemical systems. The equation is also useful for estimating the pH of a buffer solution and finding the equilibrium pH in acid-base reactions.

Two equivalent forms of the equation are

$$pH = pKa + \log_{10}\left[\frac{A}{HA}\right]$$

Or

$$pH = pK_a + \log_{10}\left[\frac{|Base|}{|Acid|}\right]$$

Here, pK_a is $-log_{10}(K_a)$ where K_a is the acid disassociation constant, that is

For the reaction:

$$HA + H2O \rightleftharpoons A^- + H_3O^+$$

In these equations, A$^-$ denotes the ionic form of the relevant acid. Bracketed quantities such as [*Base*] and [*Acid*] denote the molar concentration of the quantity enclosed

Solution	pH	litmus test	Is the solution acidic or basic?
	pink paper	blue paper	
tap water			
lemon juice			
cola soft drink			
detergent/soap*			
household ammonia			

What would be the resulting pH if you mix equal amounts of an acid and a base that are the same numerical distance from neutrality (e.g., pH 5 + pH 9)

INTERFERENCES

The glass electrode is relatively free from interference from color, turbidity, colloidal matter, oxidants, reductants, or high salinity, except for a sodium error at pH > 10. Reduce this error by using special "low sodium error" electrodes.

pH measurements are affected by temperature in two ways: mechanical effects that are caused by changes in the properties of the electrodes and chemical effects caused by equilibrium changes. In the first instance, the Nernstian slope increases with increasing temperature and electrodes take time to achieve thermal equilibrium. This can cause long-term drift in pH. Because chemical equilibrium affects pH, standard pH buffers have a specified pH at indicated temperatures.

Always report temperature at which pH is measured.

pH METER

Consist of potentiometer, a glass electrode, a reference electrode, and a temperature-compensating device. A circuit is completed through the potentiometer when the electrodes are immersed in the test solution. Many pH meters are capable of reading pH or millivolts and some have scale expansion that permits reading to 0. 001 pH unit, but most instruments are not that precise.

For routine work use a pH meter accurate and reproducible to 0.1 pH unit with a range of 0 to 14 and equipped with a temperature-compensation adjustment.

Although manufacturers provide operating instructions, the use of different descriptive terms may be confusing. For most instruments, there are two controls: intercept (set buffer, asymmetry, standardize) and slope (temperature, offset); their functions are shown diagramatically in Figures. The intercept control shifts the response curve laterally to pass through the isopotential point with no change in slope. This permits bringing the instrument on scale (0 mV) with a pH 7 buffer that has no change in potential with temperature.

REFERENCE ELECTRODE

Reference electrode consisting of a half cell that provides a constant electrode potential, commonly used are calomel and silver: silver-chloride electrodes. Either is available with several types of liquid junctions.

The liquid junction of the reference electrode is critical because at this point the electrode forms a salt bridge with the sample or buffer and a liquid junction potential is generated that in turn affects the potential produced by the reference electrode. Reference electrode junctions may be annular ceramic, quartz, or asbestos fiber, or the sleeve type. The quartz type is most widely used. The asbestos fiber type is not recommended for strongly basic solutions. Follow the manufacturer's recommendation on use and care of the reference electrode.

Refill non sealed electrodes with the correct electrolyte to proper level and make sure junction is properly wetted.

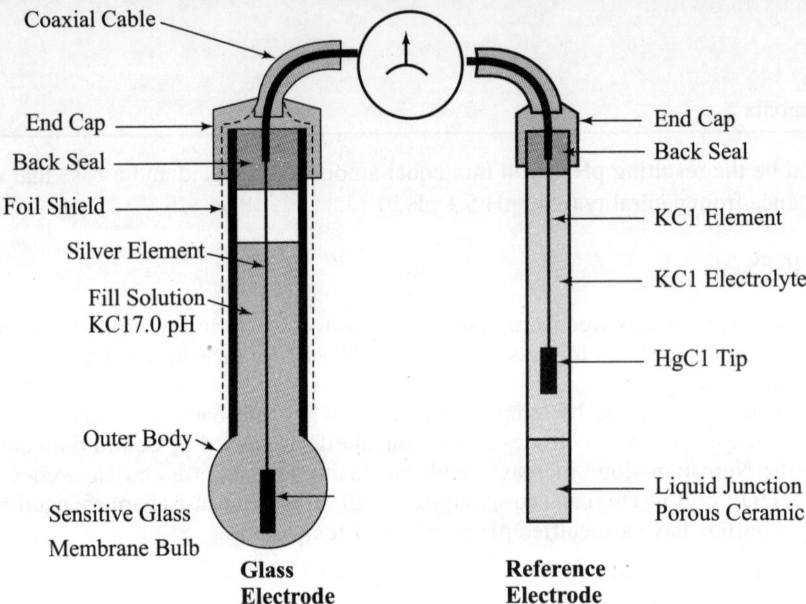

Fig. 23   Structure and components of glass electrodes

GLASS ELECTRODE

The sensor electrode is a bulb of special glass containing a fixed concentration of HCl or a buffered chloride solution in contact with an internal reference electrode. Upon immersion of a new electrode in a solution the outer bulb surface becomes hydrated and exchanges sodium ions for hydrogen ions to build up a surface layer of hydrogen ions. This, together with the repulsion of anions by fixed, negatively charged silicate sites, produces at the glass-solution interface a potential that is a function of hydrogen ion activity in solution.

Several types of glass electrodes are available. Combination electrodes incorporate the glass and reference electrodes into a single probe. Use a "low sodium error" electrode that can operate at high temperatures for measuring pH over 10 because standard glass electrodes

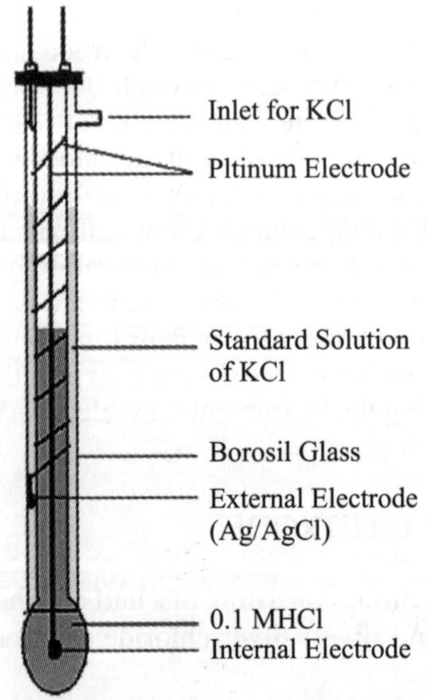

Fig. 24   Structure of a combined pH electrode

yield erroneously low values. For measuring pH below 1 standard glass electrodes yield erroneously high values; use liquid membrane electrodes instead.

BEAKERS

Preferably use polyethylene or Teflon or equivalent beakers.

STIRRER

Use either a magnetic, TFE-coated stirring bar or a mechanical stirrer with inert plastic-coated impeller.

FLOW CHAMBER

Use for continuous flow measurements or for poorly buffered solutions.

Note:

(1) Because the concentration of HCl in the glass electrode changes by repeated use, the pH meter should be calibrated against a standard solution of known pH every time.

(2) A 0.05 M solution of potassium hydrogen phthalate at 15°C has a pH of 4.000. The pH at other temperatures ($0° C - 60° C$) can be obtained from the following equation,

$$pH = 4.000 + 0.5\ \frac{(t-15)^2}{100}\ ,\text{where 't' is the temperature of the solution.}$$

(3) Ordinary glass electrode can give a linear response up to pH 13. These electrodes are not suitable for use in very strongly alkaline solutions. In such cases, membranes made of special glass, containing alkaline earth or lithium is used.

The pH of a buffer not only on the concentration of buffer ions but also depends on the activity co-efficient of the ions. This parameter is strongly affected by the total concentrations of ions in the solution (actually the ionic strength). Hence, the pH of a buffer will vary both with its own concentration and also with the concentration of other salts in the solution.

REAGENTS

General preparation

Calibrate the electrode system against standard buffer solutions of known pH. Because buffer solutions may deteriorate as a result of mold growth or contamination, prepare fresh as needed for accurate work by weighing the amounts of chemicals specified in Table 4500-H$^+$:I, dissolving in distilled water at 25°C, and diluting to 1000 mL. This is particularly important for borate and carbonate buffers.

Boil and cool distilled water having a conductivity of less than 2 µmhos/cm. To 50 mL add 1 drop of saturated KCl solution suitable for reference electrode use. If the pH of this test solution is between 6.0 and 7.0, use it to prepare all standard solutions.

Dry KH_2PO_4 at 110 to 130°C for 2 h before weighing but do not heat unstable hydrated potassium tetroxalate above 60°C nor dry the other specified buffer salts.

Although ACS-grade chemicals generally are satisfactory for preparing buffer solutions, use certified materials available from the National Institute of Standards and Technology when the greatest accuracy is required. For routine analysis, use commercially available buffer tablets, powders, or solutions of tested quality. In preparing buffer solutions from solid salts, insure complete solution.

As a rule, select and prepare buffer solutions classed as primary standards in Table 4500-H^+:I; reserve secondary standards for extreme situations encountered in wastewater measurements. Consult Table 4500-H^+:II for accepted pH of standard buffer solutions at temperatures other than 25°C. In routine use, store buffer solutions and samples in polyethylene bottles. Replace buffer solutions every 4 weeks.

SATURATED POTASSIUM HYDROGEN TARTRATE SOLUTION

Shake vigorously an excess (5 to 10 g) of finely crystalline $KHC_4H_4O_6$ with 100 to 300 mL distilled water at 25°C in a glass-stoppered bottle. Separate clear solution from undissolved material by decantation or filtration. Preserve for 2 months or more by adding one thymol crystal (8 mm diam) per 200 mL solution.

SATURATED CALCIUM HYDROXIDE SOLUTION

Calcine a well-washed, low-alkali grade $CaCO_3$ in a platinum dish by igniting for 1 h at 1000°C. Cool, hydrate by slowly adding distilled water with stirring, and heat to boiling. Cool, filter, and collect solid $Ca(OH)_2$ on a fritted glass filter of medium porosity. Dry at 110°C, cool, and pulverize to uniformly fine granules. Vigorously shake an excess of fine granules with distilled water in a stoppered polyethylene bottle. Let temperature come to 25°C after mixing. Filter supernatant under suction through a sintered glass filter of medium porosity and use filtrate as the buffer solution. Discard buffer solution when atmospheric CO_2 causes turbidity to appear.

PREPARATION OF pH STANDARD SOLUTIONS

		Weight of Chemicals Needed/ 1000 mL
Standard Solution (molality)p	pH at 25°C	Aqueous Solution at 25°C
Primary Standards		
Potassium hydrogen tartrate (saturated at 25°C)	3.557	> 7 g $KHC_4H_4O_6$*
0.05 potassium dihydrogen citrate	3.776	11.41 g $KH_2C_6H_5C_7$
0.05 potassium hydrogen phthalate	4.004	10.12 g $KHC_8H_4O_4$
0.025 potassium dihydrogen phosphate + 0.025 disodium hydrogen phosphate	6.863	3.387 g KH_2PO_4 + 3.533 g Na_2HPO_4†
0.008 695 potassium dihydrogen phosphate + 0.030 43 disodium hydrogen phosphate	7.415	1.179 g KH_2PO_4 + 4.303 g Na_2HPO_4†
0.01 sodium borate decahydrate (borax)	9.183	3.80 g $Na_2B_4O_7 10H_2O$†
0.025 sodium bicarbonate + 0.025 sodium carbonate	10.014	2.092 g $NaHCO_3$ + 2.64 g Na_2CO_3
Secondary Standards		
0.05 potassium tetroxalate dihydrate	1.679	12.61 g $KH_3C_4O_8 2H_2O$
Calcium hydroxide (saturated at 25°C)	12.454	> 2 g $Ca(OH)_2$*

*Approximate solubility.

†Prepare with freshly boiled and cooled distilled water (carbon-dioxide-free).

PROCEDURE

Instrument calibration

In each case follow manufacturer's instructions for pH meter and for storage and preparation of electrodes for use. Recommended solutions for short-term storage of electrodes vary with type of electrode and manufacturer, but generally have conductivity greater than 4000 ìmhos/cm. Tap water is a better substitute than distilled water, but pH 4 buffer is best for the single glass electrode and saturated KCl is preferred for a calomel and Ag/AgCl reference electrode. Saturated KCl is the preferred solution for a combination electrode. Keep electrodes wet by returning them to storage solution whenever pH meter is not in use.

Before use, remove electrodes from storage solution, rinse, blot dry with a soft tissue, place in initial buffer solution, and set the isopotential point. Select a second buffer within 2 pH units of sample pH and bring sample and buffer to same temperature, which may be the room temperature, a fixed temperature such as 25°C, or the temperature of a fresh sample. Remove electrodes from first buffer, rinse thoroughly with distilled water, blot dry, and immerse in second buffer. Record temperature of measurement and adjust temperature dial on meter so that meter indicates pH value of buffer at test temperature (this is a slope adjustment).

Use the pH value listed in the tables for the buffer used at the test temperature. Remove electrodes from second buffer, rinse thoroughly with distilled water and dry electrodes as indicated above. Immerse in a third buffer below pH 10, approximately 3 pH units different from the second; the reading should be within 0.1 unit for the pH of the third buffer. If the meter response shows a difference greater than 0.1 pH unit from expected value, look for trouble with the electrodes or potentiometer.

The purpose of standardization is to adjust the response of the glass electrode to the instrument. When frequent measurements are made and the instrument is stable, standardize less frequently. If sample pH values vary widely, standardize for each sample with a buffer having a pH within 1 to 2 pH units of the sample.

SAMPLE ANALYSIS

Establish equilibrium between electrodes and sample by stirring sample to insure homogeneity; stir gently to minimize carbon dioxide entrainment. For buffered samples or those of high ionic strength, condition electrodes after cleaning by dipping them into sample for 1 min. Blot dry, immerse in a fresh portion of the same sample, and read pH.

With dilute, poorly buffered solutions, equilibrate electrodes by immersing in three or four successive portions of sample. Take a fresh sample to measure pH.

WORKING OF A pH METER

(a) Soak the electrode in distilled water of pH 6-7 for 24 hours before using it.
(b) Fix the 'temperature compensation knob' to the temperature of the samples are to be measured. This is important because the equilibrium constant of a solution vary with temperature.
(c) The electrode should be rinsed with distilled water and wiped gently with tissue paper. The electrode should be gently immersed into a standard buffer solution with 7 pH. The electrode should not touch the sides or bottom of the breaker. Adjust the pH meter to the standard pH, using the "calibration knob".
(d) Once calibrated, rinse the electrodes with distilled water and wipe it gently with tissue paper.

(e) Dip the electrode now into the solution, whose pH should be found out. Note the pH of the test solution.

(f) Wash and store the electrode, after the completion of the experiment, in distilled water.

TROUBLE SHOOTING

Electrodes

If potentiometer is functioning properly, look for the instrument fault in the electrode pair. Substitute one electrode at a time and cross-check with two buffers that are about 4 pH units apart. A deviation greater than 0.1 pH unit indicates a faulty electrode. Glass electrodes fail because of scratches, deterioration, or accumulation of debris on the glass surface. Rejuvenate electrode by alternately immersing it three times each in 0.1N HCl and 0.1N NaOH. If this fails, immerse tip in KF solution for 30 s. After rejuvenation, soak in pH 7.0 buffer overnight. Rinse and store in pH 7.0 buffer. Rinse again with distilled water before use. Protein coatings can be removed by soaking glass electrodes in a 10% pepsin solution adjusted to pH 1 to 2.

To check reference electrode, oppose the emf of a questionable reference electrode against another one of the same type that is known to be good. Using an adapter, plug good reference electrode into glass electrode jack of potentiometer; then plug questioned electrode into reference electrode jack. Set meter to read millivolts and take readings with both electrodes immersed in the same electrolyte (KCl) solution and then in the same buffer solution. The millivolt readings should be 0 ± 5 mV for both solutions. If different electrodes are used, i.e., silver: silver-chloride against calomel or vice versa, the reading will be 44 ± 5 mV for a good reference electrode.

Reference electrode troubles generally are traceable to a clogged junction. Interruption of the continuous trickle of electrolyte through the junction causes increase in response time and drift in reading. Clear a clogged junction by applying suction to the tip or by boiling tip in distilled water until the electrolyte flows freely when suction is applied to tip or pressure is applied to the fill hole. Replaceable junctions are available commercially.

Problem 1: What is the pH of a solution with a volume of 475 mL containing a total of 1.20 grams of hydrochloric acid?

Problem 2: What's the pH of a 0.750 M formic acid (HCO_2H) solution? K_a (HCO_2H) $= 1.77 \times 10^{-4}$ M.

- pH = $-\log[H^+]$
- pH = $-\log(0.00296)$
- pH = 2.53

Problem 3: What's the pH of a 0.00340 M NaOH solution?

- K_w = $[H^+][OH^-]$
- $1.00 \times 10^{-14} = [H^+][0.0500]$
- $[H^+]$ = 2.00×10^{-13} M

To find pH, we simply use the equation:

- pH = $-\log[H^+]$
- pH = $-\log[2.00 \times 10^{-13}]$
- pH = 12.7

This answer indicates a very basic solution, which is what we would expect from a 0.0500 M NaOH solution.

Problem 4: If blood had a normal pH of 6.1 instead of 7.2, would you expect exercise to result in heavy breathing? Justify your answer.

36

BIOLOGICAL BUFFERS

INTRODUCTION

A buffer, as defined is "a substance which by its presence in solution increases the amount of acid or alkali that must be added to cause unit change in pH". Buffers are thus very important components in experiments designed to study biological reactions by maintaining a constant concentration of hydrogen ions within the physiological range. The pH of mammalian blood is maintained close to 7.38 by buffer systems such as

$$H_2PO_4 <=> HPO_4^{2}, CO_2 <=> H_2CO_3,$$

$$H_2CO_3 <=> HCO_3^{-},$$

many organic acids, organic bases and proteins. In living plants, the normal range of pH in tissues is about 4.0-6.2. It is not as narrowly defined as in mammalian tissues.

Universally applicable buffers for biochemistry must display:
- water solubility
- no interference with biological processes
- known complex-forming tendency with metal ions
- non-toxicity
- no interference with biological membranes (penetration, solubilisation, adsorption on surface etc.)

PRIMARY STANDARD BUFFERS

Composition and properties of the five primary standard buffers at 25°C

BUFFER SOLUTION

	Tartrate	Phthalate	Phosphate D	Phosphate E	Borate
Buffer substance	$KHC_4H_4O_6$	$KHC_8H_4O_4$	KH_2PO_4 Na_2HPO_4	+ KH_2PO_4 Na_2HPO_4	+ $Na_2B_4O_7$ 10 H_2O
g/l soln. at 25 °C	Saturated at 25°C	10.12 3.53 [c]	3.39 4.30 [c]	[b] 1.179	[b] 3.80°C

Cont...

Cont...

Molality (m)	0.0341	0.05	0.025 [a] 0.03043 [c]	0.008695	[b] 0.01
Molarity (M)	0.034	0.04958	0.02490 [a] 0.03032 [c]	0.008665 [b]	0.009971
Density (g/ml)	1.0036	1.0017	1.0028	1.0020	0.9996
pH at 25°C	3.557	4.008	6.865	7.413	9.180
Dilution value, D pH$^1/_2$	+0.049	+0.052	+0.080	+0.07 [d]	+0.01
Buffer value, equiv./pH	b, 0.027	0.016	0.029	0.016	0.020
Temp. coeff., dpH(S)/dt, units/°C	−0.0014	+0.0012	−0.0028	−0.0028	−0.0082

One of the most important applications of acids and bases in chemistry and biology is that of buffers. A buffer solution resists rapid changes in pH when acids and bases are added to it. Every living cell contains natural buffer systems to maintain the constant pH needed for proper cell function. Many consumer products are also buffered to safeguard their activity. What are buffers made of? How do buffers maintain the delicate pH balance needed for life and health?

The ability of buffers to resist changes in pH when acid or base is added is a result of their chemical composition. All buffers contain a mixture of a conjugate acid–base pair; either a weak acid (HA) and its conjugate base (A–), or a weak base (B), and its conjugate acid (BH+). Weak acids and weak bases both dissociate slightly in water (Reactions 1 and 2).

$$\text{Ka } HA(aq) + H2O(l) \rightleftharpoons H3O+(aq) + A–(aq) \quad \text{Reaction 1}$$

weak acid conjugate base

$$\text{Kb } B(aq) + H2O(l) \rightleftharpoons BH+(aq) + OH–(aq) \quad \text{Reaction 2}$$

weak base conjugate acid

These reactions are reversible and both the weak acid and its conjugate base or the weak base and its conjugate acid are present in solution.

The equilibrium constant expressions for these dissociation reactions are:

$$\text{Ka} = [H3\,O+][A–]/[HA] \quad \text{Equation 1}$$

$$\text{Kb} = [OH–][BH+]/[B] \quad \text{Equation 2}$$

Buffers control pH because the two buffering components, either HA and A– or B and BH+, are able to neutralize both acids and bases added to the solution.

$$HA(aq) + OH^-\,(aq) \longrightarrow H2O(l) + A^-(aq) \quad \text{Reaction 3}$$

$$A^-(aq) + H3O^+(aq) \longrightarrow H2O(l) + HA(aq) \quad \text{Reaction 4}$$

$$BH^+(aq) + OH^-(aq) \longrightarrow H2O(l) + B(aq) \quad \text{Reaction 5}$$

$$B(aq) + H3O^+(aq) \longrightarrow H2O(l) + BH^+(aq) \quad \text{Reaction 6}$$

The actual pH of a buffer solution depends on the concentration of the conjugate acid–base pair in solution. If Equation 1 is rearranged, the concentration of hydronium ions in solution is:

$$[H_3O+] = K_a \times \frac{[HA]}{[A-]} \qquad\qquad \text{Equation 3}$$

and the pH is:

$$pH = -\log[H_3O+] = pK_a - \log\frac{[HA]}{[A-]} \qquad\qquad \text{Equation 4}$$

If the concentrations of the acid–base pair are equal, [HA] = [A–]. The

$-\log\frac{[HA]}{[A-]}$ is equal to zero, and the pH of the buffer is equal to pKa. By varying the amounts

HA and A– in solution, the pH of the buffer solution can be changed.

For a buffer made up of a weak base (B) and its conjuugate acid (BH+), the solution pH calculations are similar. If Equation 2 is rearranged, the concentration of hydroxide ions (OH–) in solution is:

$$[OH-] = K_b \times \frac{[B]}{[BH+]} \qquad\qquad \text{Equation 5}$$

and the pOH is:

$$pOH = -\log[OH-] = pK_b - \log\frac{[B]}{[BH+]} \qquad\qquad \text{Equation 6}$$

If pOH is known, then pH can be calculated using Equation 7:

pH + pOH = 14.00

pH = 14.00 – pOH $\qquad\qquad$ Equation 7

Once the buffer is made, how does the pH remain constant when strong acid or base is added? Acetic acid is a weak acid. If a buffer solution is made with 0.5 moles of acetic acid and 0.5 moles of its conjugate base sodium acetate, the initial pH of the solution will be equal to pKa, or 4.74. Now, if 0.05 moles of a strong acid is added to the buffer, the H3O+ will react with 0.05 moles of the sodium acetate to form 0.05 moles of acetic acid. This produces a solution with 0.55 moles of acetic acid and 0.45 moles of its conjugate base sodium acetate. If the solution volume change is slight, then the new pH of the solution is:

$$pH = 4.74 - \log\frac{0.55}{0.45} = 4.65$$

The pH difference is only 0.09 units!

For buffers to be effective, noticeable amounts of both the conjugate acid–base pair must be present in solution. This limits the concentration ratios for HA:A– or B:BH+ to between 10:1 and 1:10 and the pH range for the buffering action of any weak acid to pKa ±1. An ideal buffer is a solution that contains equal numbers of moles of the conjugate acid–base pair.

PROBLEMS

1. How many grams of sodium acetate (molar mass 82.03 g/mol) must be added to 1.00 L of a 0.200 M acetic acid solution to form a buffer of 4.20? Ka value for acetic acid is 1.8×10^{-5}.

2. Three milliliters of a 2.0 M solution of HCl are added to 1 liter of buffer solution containing 0.40 moles of the weak acid, propanoic acid ($Ka = 1.4 \times 10^{-5}$) and 0.50 moles of its conjugate base, sodium propanate.

   a. What is the original pH of the buffer before the strong acid is added?
   b. What is the pH of the buffer after the HCl is added? Assume negligible volume change.

3. The weak base–conjugate acid buffer used in this laboratory consists of a weak base ammonia, NH_3, and its conjugate acid ammonium chloride, NH_4Cl. If the NH_3 concentration is 0.05 M and the NH4Cl concentration is 0.05 M, what is the pH of the buffer? Kb for NH3 is 1.8×10^{-5}.

BUFFERING PROPERTIES OF A WEAK ACID–CONJUGATE BASE BUFFER

1. Obtain 25.0 mL of the 0.1 M acetic acid solution in a 25-mL graduated cylinder. Transfer the acetic acid solution to a clean 100-mL beaker.
2. Rinse the 25-mL graduated cylinder with deionized water. Obtain 25.0 mL of the 0.1 M sodium acetate solution with the 25-mL graduated cylinder.
3. Transfer the sodium acetate solution to the 100-mL beaker containing 25 mL of the acetic acid solution. Stir the solution.
4. Set-up a pH meter and electrode (or a pH sensor). Calibrate the pH meter using a standard pH 7 buffer solution.
5. Remove the pH 7 buffer solution, place a 100-mL beaker under the electrode, and rinse the electrode well with deionized water.
6. Set the 100-mL beaker containing the acetic acid–acetate buffer solution on a magnetic stirrer, if one is available. Add a stir bar to the solution. Gently stir the buffer solution.
7. Place the pH electrode in the solution. Record the pH of the solution in the Part 1 Data Table. (0 mL of 0.2 M HCl added.)
8. Obtain approximately 30 mL of 0.2 M HCl solution in a clean 50-mL beaker and label the beaker, 0.2 HCl.
9. Using a graduated Beral-type pipet, transfer 1.0 mL of the 0.2 M HCl solution to the acetic acid–acetate buffer solution.
10. Record the pH of the solution in the Part 1 Data Table. (1 mL of 0.2 M HCl added)
11. Using the same graduated pipet, transfer another 1.0 mL of the 0.2 M HCl solution to the acetic acid–acetate buffer solution.
12. Record the pH of the solution.
13. Repeat steps 11 and 12, recording the pH in the Part 1 Data Table after each 1.0 mL of HCl is added, until a total of 10.0 mL of HCl has been added to the solution.
14. Remove the electrode from the solution, place a 100-mL beaker under the electrode, and rinse the electrode with deionized water.
15. Dispose of the acetic acid–sodium acetate solution as directed by the instructor, and rinse the 100-mL beaker with deionized water. Do not dispose of the HCl.
16. Obtain 25.0 mL of the 0.1 M acetic acid solution in a clean 25-mL graduated cylinder. Transfer the acetic acid solution to the rinsed 100-mL beaker.
17. Rinse the 25-mL graduated cylinder with deionized water. Obtain 25.0 mL of the 0.1 M sodium acetate solution in the 25-mL graduated cylinder.
18. Transfer the sodium acetate solution to the 100-mL beaker containing 25 mL of the acetic acid solution. Stir the solution.

19. Place the 100-mL beaker on the magnetic stirrer. Add a stir bar and gently stir the buffer solution.
20. Place the pH electrode in the solution. Record the pH of the solution in the data table. (0 mL of 0.2 M NaOH added.)
21. Obtain approximately 30 mL of 0.2 M NaOH solution in a clean 50-mL beaker. Label the beaker, 0.2 M NaOH.
22. Using a clean graduated Beral-type pipet, transfer 1.0 mL of the 0.2 M NaOH solution to the acetic acid–acetate buffer solution.
23. Record the pH of the solution.
24. Repeat steps 22 and 23, recording the pH in the data table after each 1.0 mL of NaOH solution is added, until a total of 10.0 mL of NaOH has been added.
25. Remove the pH electrode from the solution, place a 100-mL beaker under the electrode, and rinse the electrode with deionized water.
26. Dispose of the acetic acid–sodium acetate solution as directed by the instructor, and rinse the 100-mL beaker with deionized water. Do not dispose of the NaOH solution.

BUFFERING PROPERTIES OF A WEAK BASE–CONJUGATE ACID BUFFER

1. Obtain 50 mL of the ammonia–ammonium chloride buffer solution in a 50-mL graduated cylinder.
2. Transfer the 50 mL of buffer solution to a clean 100-mL beaker.
3. Set the 100-mL beaker on the magnetic stirrer. Add a stir bar to the solution. Gently stir the buffer solution.
4. Place the pH electrode in the solution. Record the pH of the solution in the Part 2 Data Table. (0 mL of 0.2 M HCl added.)
5. Using a clean graduated Beral-type pipet, transfer 1.0 mL of 0.2 M HCl solution to the ammonia–ammonium chloride buffer solution. Record the pH .
6. Repeat step 5, recording the pH in the Part 2 Data Table after each 1.0 mL of HCl is added, until a total of 10.0 mL of HCl has been added.
7. Remove the electrode from the solution, place a 100-mL beaker under the electrode and rinse the electrode with deionized water.
8. Dispose of the ammonia–ammonium chloride solution as directed by the instructor and rinse the 100-mL beaker with deionized water.
9. Obtain another 50 mL of the ammonia–ammonium chloride buffer solution in a clean 50-mL graduated cylinder.
10. Transfer the 50 mL of buffer solution to a clean 100-mL beaker.
11. Set the beaker on a magnetic stirrer. Add a stir bar to the beaker and gently stir the buffer solution.
12. Place the pH electrode in the solution. Record the pH of the solution. (0 mL of 0.2 M NaOH added.)
13. Using a clean graduated Beral-type pipet, transfer 1.0 mL of the 0.2 M NaOH solution to the ammonia–ammonium chloride buffer solution. Record the pH.
14. Repeat step 13, recording the pH in the Part 2 Data Table after each 1.0 mL of NaOH is added, until a total of 10.0 mL of NaOH has been added.
15. Remove the electrode from the solution, place a 100-mL beaker under the electrode, and rinse the electrode with deionized water.
16. Dispose of the ammonia–ammonium chloride solution as directed by the instructor.

PREPARATION OF A pH 5.00 BUFFER SOLUTION

1. Calculate the correct volumes of the 0.1 M acetic acid solution and the 0.1 M sodium acetate solution needed to make 50 mL of a buffer solution with a pH value of 5.00. Record these amounts in the data table. Ka = 1.8 x 10^{-5}.
2. Obtain the calculated volume of 0.1 M acetic acid solution in a clean 50-mL graduated cylinder and transfer this volume to a clean 100-mL beaker.
3. Rinse the graduated cylinder with deionized water. Obtain the calculated volume of 0.1 M sodium acetate in the graduated cylinder and transfer the volume to the 100-mL beaker containing the acetic acid solution. Stir the solution.
4. Place the beaker under the pH electrode.
5. Place the pH electrode in the acetic acid–soidium acetate buffer solution. Record the pH.
6. Remove the electrode from the solution, place a 100-mL beaker under the electrode, and rinse the electrode with deionized water.
7. Dispose of the acetic acid–acetate buffer solution as directed by the instructor.
8. Store and put away the pH electrode as directed by the instructor.

pH of Acetic Acid- Sodium Acetate Buffer

mL. of 0.2 M	pH		mL. of 2M	pH	
HCl added	actual	calc.	NaOH added	actual	calc.
0			0		
1.0			1.0		
2.0			2.0		
3.0			3.0		
4.0			4.0		
5.0			5.0		
6.0			6.0		
7.0			7.0		
8.0			8.0		
9.0			9.0		
10.0			10.0		

mL of 0.1 M CH_3COOH ___________ mL pH ___________ (calc.)
mL of 0.1 M $NaCH_3COO$ ___________ mL pH ___________ (actual)

pH of *Ammonia–Ammonium Chloride Buffer*

mL. of 0.2 M HCl added	pH actual	calc.	mL. of 2M NaOH added	pH actual	calc.
0			0		
1.0			1.0		
2.0			2.0		
3.0			3.0		
4.0			4.0		
5.0			5.0		
6.0			6.0		
7.0			7.0		
8.0			8.0		
9.0			9.0		
10.0			10.0		

PROBLEMS

1. Using Equation 4 on page 2, calculate the pH of the Part 1 acetic acid–sodium acetate buffer solution before and after 1.0 mL of 0.2 M HCl solution is added to the buffer. K_a of acetic acid equals 1.8×10^{-5}. Enter the values in the data table.

2. When strong base is added to a buffer of a weak acid–conjugate base, the acid reacts with the base to form water and its conjugate base.

$$HA(aq) + OH^-(aq) \rightleftharpoons H_2O(l) + A^-(aq)$$

Calculate the pH of the Part 1 acetic acid–sodium acetate buffer solution after 1.0 mL of the 0.2 M NaOH solution is added to the buffer. Enter this value in the Data Table.

3. Repeat the pH calculation for each successive 1.0 mL increment of 0.2 M NaOH added to the buffer. Enter these values in the Data Table.

4. The ammonia–ammonium chloride buffer solution is a weak base–conjugate acid buffer solution.

K_b for NH3 equals 1.8×10^{-5}. Using Equation 4

$$pH = 14.0 - pOH$$

Calculate the pH of the ammonia–ammonium chloride buffer solution after 1.0 mL of 0.2 M HCl is added to the buffer solution. The initial moles of both NH3 and NH4Cl in 50 mL of the buffer solution are 0.0025 moles. Record the pH value in the Part 2 Data Table.

$$[NH_3(aq) + H_3O+(aq) \rightleftharpoons NH_4(aq) + H_2O(l).]$$

5. Calculate the pH change when of 1 mL of 0.2 M HCl is added to 50 mL of deionized water. How does this pH value change compare to those obtained when 1 mL of 0.2 M HCl is added to the buffers?

37

CENTRIFUGATION

CENTRIFUGATION

T. Svedberg and J.W. Williams in the 1920's accomplished the single most important advance in biomedical instrumentation by the use of centrifugal force to separate biologically important substances by coupling of mechanics, optics and mathematics. In honor of that work, the value for a molecules (or organelle's) sedimentation velocity in a centrifugal field is known as its Svedberg constant or S value for short.

A centrifuge is a device for separating particles from a solution according to their size, shape, density, viscosity of the medium and rotor speed. In biology, the particles are usually cells, sub cellular organelles, viruses, large molecules such as proteins and nucleic acids. There are many ways to classify centrifugation.

Two principal uses

1. Separate out solid matter as a Pellet from dissolved solutes as Supernatant
2. Separate soluble macromolecules of different mass or density on the basis of size, shape or density.

Also sometimes used to provide centrifugal force to drive other processes,eg ultrafiltration. A centrifuge is used to separate particles or even macromolecules: Cells, Sub-cellular components, Proteins, Nucleic acids

METHODOLOGY

- Utilizes density difference between the particles/macromolecules and the medium in which these are dispersed
- Dispersed systems are subjected to artificially induced gravitational fields

PRINCIPLES OF CENTRIFUGATION

Centrifugation is a method for separating particles from each other in a solution. In biology these particles are usually cells, subcellular organelles or large molecules. The basic principles of sedimentation originate from Stoke's Law which describes the settling of a sphere in a gravitational field.

Stokes Law states that for spherical particles:

$$u = \frac{gd^2\partial\rho}{18\,\eta}$$

Where:

- u is the terminal upward velocity of the inclusion
- g is 9.81 m s^{-1}
- d is the inclusion diameter [m]
- $\partial\rho$ is the difference in densities between the liquid steel and inclusion [kg m^{-3}], and
- η is the molten steel viscosity [N s m^{-2}]

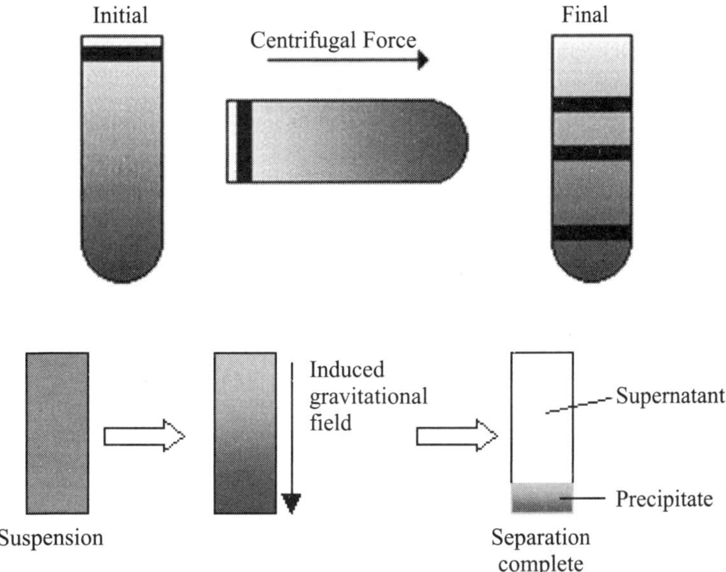

Fig. 25   Mechanism of centrifugation

From this arrangement of the Stoke's Law equation, it can be seen that:

- The sedimentation velocity is proportional to the square of the sphere size.
- The sedimentation velocity is proportional to the difference between the particle and liquid density and is zero when these two densities are equal.
- The sedimentation velocity decreases as the viscosity of the liquid medium increases.
- The sedimentation velocity increases as the centrifugal force increases.

Sedimenting force on particle = Mass × centrifugal field = $\mathbf{mw^2r}$

where w = angular velocity of rotor (radians/sec)

r = radius (*ie* distance of particle from axis of rotation)

Relative Centrifugal Force (RCF)

$$RCF = 1.119 \times 10^{-5} \times (rpm)^2 \times r$$

RCF value reported as "No. × g" (ie multiples of earth's gravitational (force).

Interacting Forces in Centrifugation

Sedimenting force, mw^2r, is opposed by...

1. **Flotation Force** (Archimedes) = mw^2rvr
   Where: v = **partial specific volume** (volume displaced by 1g of sedimenting particles)

$$r = \text{density of solution}$$

Net Sedimentating Force on particle, **after allowing for flotation**

$$= mw^2r\ (1 - vr)$$

2. **Frictional Resistance** against particle moving through fluid $= f.v$

$$\text{where: } f = \text{frictional coefficient}$$

$$v = \text{particle velocity}$$

3. **Diffusion** - acting to counter uneven concentration distributions set up when dissolved molecules sediment.

Sedimenting force on particle $= \text{Mass} \times \text{centrifugal field} = mw^2r$

$$\text{where } w = \text{angular velocity of rotor (radians/sec)}$$

$$r = \text{radius (}ie\text{ distance of particle from axis of rotation)}$$

$$\text{Relative Centrifugal Force (RCF)}$$

$$\text{RCF} = 1.119 \times 10^{-5} \times (\text{rpm})^2 \times r$$

RCF value reported as "No. $\times$ g" (ie multiples of earth's gravitational (force).

$$\text{Interacting Forces in Centrifugation}$$

$$\text{Sedimenting force, } mw^2r, \text{ is opposed by...}$$

4. **Flotation Force** (Archimedes) = mw^2rvr

$$\text{where: } v = \textbf{partial specific volume}$$

$$\text{(volume displaced by 1g of sedimenting particles)}$$

$$r = \text{density of solution}$$

Net Sedimenting Force on particle, **after allowing for flotation** $= mw^2r(1 - vr)$

5. **Frictional Resistance**

$$\text{against particle moving through fluid} = f.v$$

$$\text{where: } f = \text{frictional coefficient}$$

$$v = \text{particle velocity}$$

6. **Diffusion-** acting to counter uneven concentration distributions set up when dissolved molecules sediment.

Balance between the Sedimentation force and Counteracting Forces leads to various formulae and equations used in Preparative Centrifugation *eg* to calculate the time required to sediment a particle to the bottom of the tube.

TYPES OF CENTRIFUGE

1. Analytical/Preparative Centrifugation

The two most common types of centrifugation are analytical and preparative; the distinction is between the two is based on the purpose of centrifugation. Analytical centrifugation involves measuring the physical properties of the sedimenting particles such as sedimentation coefficient or molecular weight. Optimal methods are used in analytical ultracentrifugation. Molecules are observed by optical system during centrifugation, to allow observation of macromolecules in solution as they move in gravitational field. The samples are centrifuged in cells having windows that lie paralleled to the plan of rotation of the rotor head. As the rotor turns, the images of the cell (proteins) are projected by an optical system on to film or a computer. The concentration of the solution at various points in the cell is determined by absorption of a light of the appropriate wavelength (Beer's law is followed). This can be accomplished either by measuring the degree of blackening of a photographic film or by the pen deflection of the recorder of the scanning system or fed into a computer.

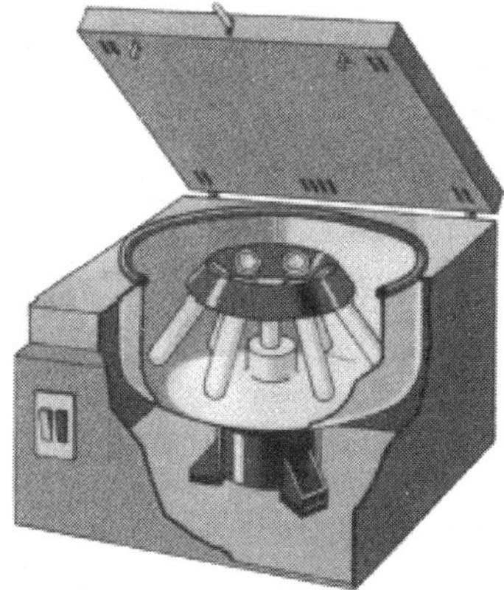

Fig. 26   A simple lab centrifuge

The other form of centrifugation is called preparative and the objective is to isolate specific particles which can be reused. There are many type of preparative centrifugation such as rate zonal, differential, isopycnic centrifugation

TYPES OF ANALYTICAL CENTRIFUGE

(a) Tubular bowl centrifuge.

This is generally operated vertically, the tubular rotor providing a long flow path enabling clarification. The sludge collects and must be removed.

(b) Continuous scroll centrifuge.

This is operated horizontally. The helical screw scrolls the solids along the bowl surface and out of the liquid; the sludge being dewatered before discharge. The clarified liquor overflows over an adjustable weir at the other end of the bowl. The screw conveyer rotates at a slightly different speed to the bowl.

(c) Continuous multichamber disc-stack centrifuge.

The bowl contains a number of parallel discs providing a large clarifying surface with a small sedimentation distance. The sludge is removed through a valve.

2. Differential Centrifugation

The most common centrifugation method involves fractionation of particles differentially between a pellet and supernatant. The pellet is a mixture of all sedimenting components while the supernatant contains purified components of only the slowest sedimenting components. This works well only for initial separations of particles with very different sedimention velocities. As can be seen at the right, the largest particles will sediment first leaving a mixture of the slower sedimenting species. Notice that yield of the lighter species is not very high since many of these particles are being trapped in the pellet.

An analytical application of this methods looks at the zone of clearing for each sedimenting species. For example, as each species sediments there will be a boundary above which there will not be any molecules of that species. Using these moving boundary methods, each sedimenting species can be monitored with respect to sedimention velocity and concentration

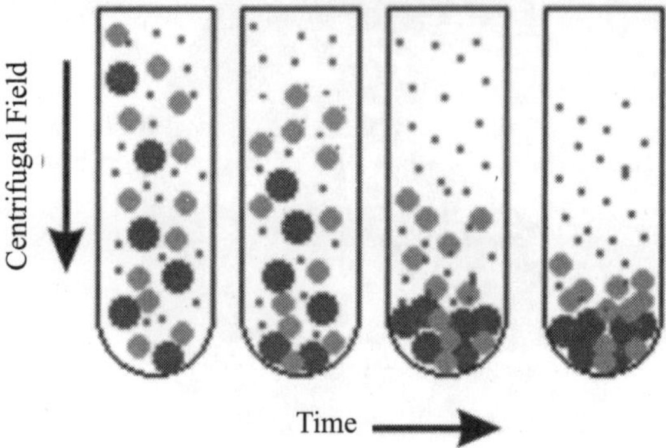

Fig. 27   Mechanism of differential centrifugation

3. Zone Velocity Centrifugation

A method that results in yields greater than differential centrifugation and allows greater resolution of all particles sizes is zone velocity centrifugation. A sample is layered on top of a shallow density gradient and then centrifuged. Each particle size will migrate as a zone or band at a characteristic velocity. If the velocities of the particles are sufficiently different then the zones of the particles will resolve. The gradient through which the particles are centrifuged is used to stablilze the zones during recovery and helps prevent mixing of resolved zones. The density of the gradient material should be less than the components being sedimented.

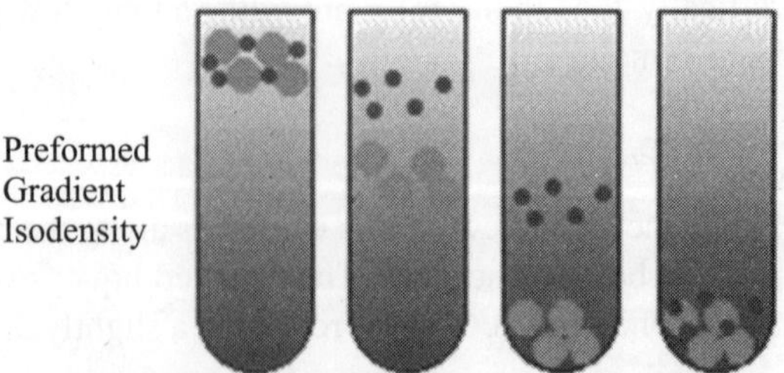

Fig. 28   Zone Velocity centrifugation

4. Density Gradient Centrifugation

In absence of a density gradient, separated **bands** of solute in the centrifuge are gravitationally unstable.

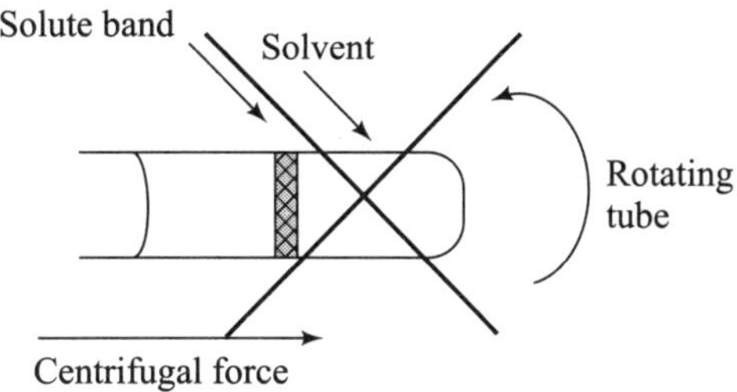

Fig. 29 Forces acting on a centrifuge tube

Separation cannot occur because layer of concentrated, dense solution overlaying less dense solvent would lead to mixing by convection and nullify the separation. In absence of stabilising density gradient, can form boundaries but not zones. In analytical ultracentrifuge, moving boundaries and concentration distributions observed by optical device.

- **Create Density Gradient in tube**
  Use a non-interacting, low M.Wt solute in continuously increasing concentration from meniscus to bottom of tube.
- Important technique for purifying **proteins** and particularly **nucleic acids**.
  Two different types of density gradient centrifugation, for two different purposes are:
- **Zonal (or Rate Zonal) Centrifugation**
  (Sucrose density gradient centrifugation)
- **Isopycnic Centrifugation**

  (Caesium chloride density gradient centrifugation)

4a. Zonal Centrifugation

Mixture to be separated is layered on top of a Sucrose or Ficol Gradient (increasing concentration down the tube) - provides gravitational stability as different species move down tube at different rates forming separate bands.

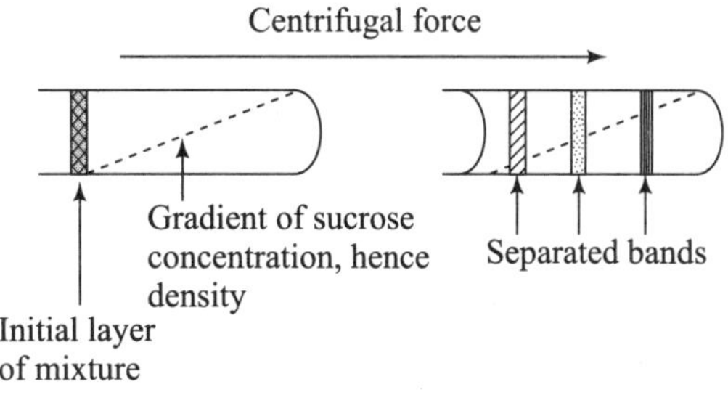

Fig. 30 Species Separated by differences in Sedimentation Coefficient (S)

$$S = \frac{\text{Rate of movement down tube}}{\text{Centrifugal force}}$$

- Sis increased for particle of Larger Masses (because sedimenting force a $M(1-vr)$
- S is also increased for More Compact Structures of equal particle mass (frictional coefficient is less)
- Mild, non-denaturing procedure, useful for protein purification, and for intact cells and organelles.

4b. Isopycnic Centrifugation

Molecules separated on equilibrium position, not by rates of sedimentation. Each molecule floats or sinks to position where density equals density of CsCl solution. Then no net sedimenting force on molecules.

Isopycnic = Equal density
and separation is on basis of different densities of the particles.

(a) Before centrifugation CsCl and sample uniformly distributed

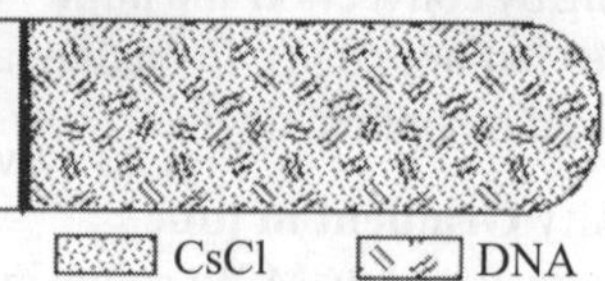

(b) After centrifugation CsCl redistributes giving density gradient. Nucleic acid species form bands at " equal density" levels

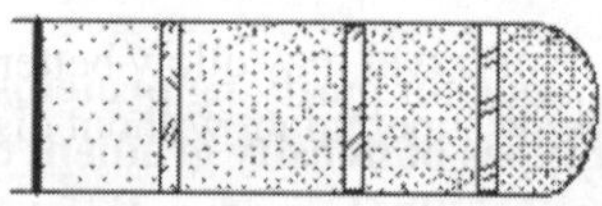

Fig. 31   CsCl gradient

Very useful for purifying nucleic acid species of different density; also in separating proteoglycans extracted from cartilage

A centrifuge is comprised of an:

- Electric motor
- Drive shaft
- Rotor to hold tubes

Sedimentation of suspended and some dissolved particles occurs due to centrifugal force.

CENTRIFUGE ROTORS

A. Fixed Angle Rotor

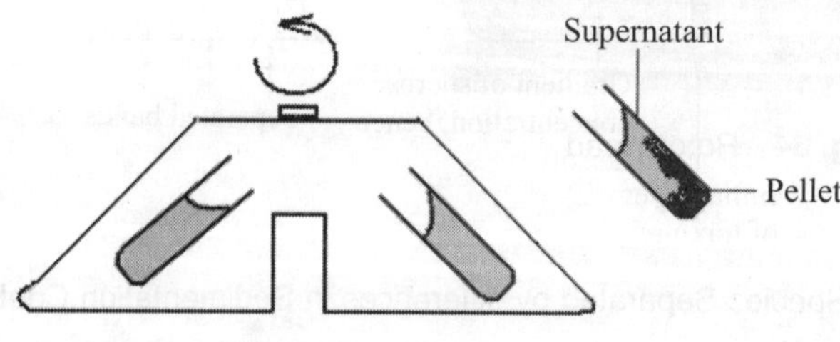

Fig. 32   Centrifuge Rotor

Advantage:

- Sedimenting particles have only short distance to travel before pelleting.
- Shorter run time.
- The most widely used rotor type

B. Swinging Bucket Rotor

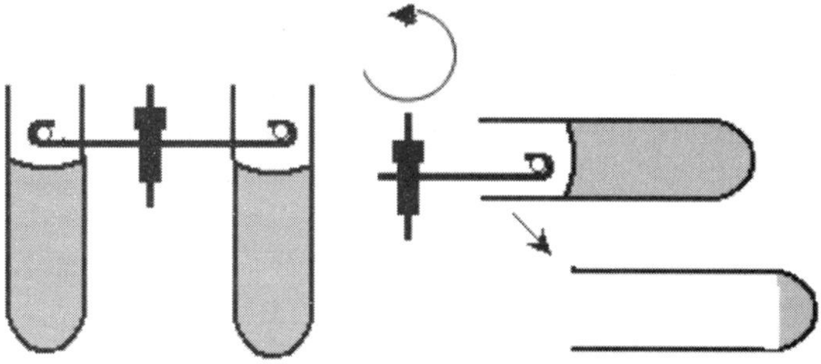

Fig. 33   Swinging Bucket Rotor

Advantage

- Longer distance of travel may allow better separation , *eg* in density gradient centrifugation.
- Easier to withdraw supernatant without disturbing pellet

A centrifuge with very light arms (neglect their mass) 10 cm long spins down samples of a colloidal mixture (one in each tube) at 3000 rpm. The arms are jointed at the top and swing out as the device spins up so that the bottom of the sample tubes exert a (normal) force on the samples that is **always directed** back along the arms.

Fig. 34   Rotor Head

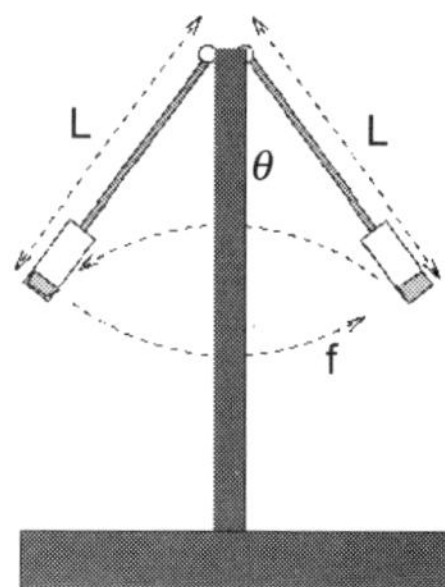

Fig. 35   Dynamics of centrifugation

Centrifuge Cautions

These cautions presume you have had proper instruction in the use of the centrifuge OR have read the instruction manual thoroughly.

1. Make sure the correct rotor is being used and that it is installed properly on the spindle.
2. Balance the load in the rotor - every tube must have a balance tube in the opposite slot with the same volume of fluid.
3. Make sure you are using the appropriate centrifuge tube for the job - they can rupture at too high a speed.
4. Pre-cool the centrifuge and the rotor. Rotors should be stored in a refrigerator.
5. Do not attempt to override any safety features of the centrifuge.
6. Never leave the centrifuge unattended until it reaches maximum speed and is going smoothly.
7. When in doubt, ask for help.

38

COLORIMETRY

COLORIMETER

The Colorimeter is a computer-interfaced probe designed to determine the concentration of a solution from its color intensity. The color of a solution may be inherent or derived by adding another reagent to it. Monochromatic light from a LED light source passes through a cuvette containing a solution sample, as shown in Figure 1. Some of the incoming light is absorbed by the solution. As a result, light of a lower intensity strikes a photodiode.

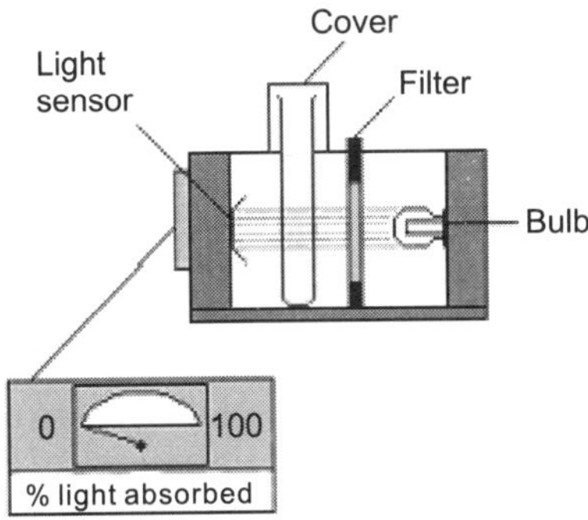

Fig. 36 Parts of a Colorimeter

TRANSMITTANCE AND ABSORBANCE

The amount of light that penetrates a solution is known as transmittance. Transmittance can be expressed as the ratio of the intensity of the transmitted light, It, and the initial intensity of the light beam, Io, as expressed by the formula:

$$T = It/Io$$

The Colorimeter produces an output voltage which varies in a linear way with transmittance, allowing a computer to monitor transmittance data for a solution. The transmittance of the sample varies logarithmically (base ten) with the product of three factors: a, the molar absorptivity (extinction coefficient), b, the cell or cuvette width, and c, the molar concentration.

$$\log(1/T) = abc$$

In addition, many experiments designed to use a colorimeter require a related measurement, absorbance. At first glance, the relationship between transmittance and absorbance would appear to be a simple inverse relationship. That is, as the amount of light transmitted by a solution increases, the amount of light absorbed might be expected to decrease proportionally. But the true relationship between these two variables is inverse and logarithmic (base 10). It can be expressed as:

$$A = \log(1/T)$$

Combining the two previous equations, the following expression is obtained:

$$A = abc$$

This formula states that the light absorbed by a solution depends on the absorbing ability of the solute, the distance traveled by the light through the solution, and the concentration of the solution. For a given solution contained in a cuvette with a constant cell width, one can assume a and b to be constant. This leads to the equation:

$$A = k\,C \text{ (Beer's law)}$$

Where, k is proportionality constant.

This equation shows absorbance to be related directly to concentration and represents a mathematical statement of Beer's law. In this guide and in our computer programs, transmittance is expressed as percent transmittance or %T. Since T= %T/100, the formula can be rewritten as:

$$A = \log (100/\%T) \text{ or } A = 2 - \log \%T$$

BEER'S LAW

In general, absorbance is important because of its direct relationship with concentration according to Beer's law. This linear relationship is obeyed only at the maximum absorbance (λ_{max}) of the compound. The λ_{max} of the compound being analyzed should correspond very closely to the available wavelengths of the LED's which are 470, 565 and 635 nm. Beer's law should be verified by using several standards (solutions of known concentration) and their absorbance values determined using a colorimeter. A graph of absorbance versus concentration is then plotted. A solution of unknown concentration is placed in the colorimeter and its absorbance measured. When the absorbance of this solution is interpolated on the Beer's law curve, as shown in the graph in figure, its concentration is determined on the horizontal axis. Alternatively, its concentration may be found using the slope of the Beer's law curve.

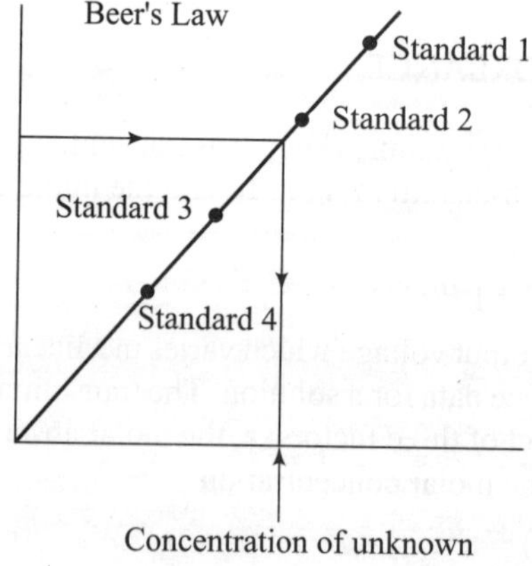

Absorbance and Transmittance Ranges for the Colorimeter

For the best results, the absorbance or transmittance values should fall within these ranges:
Percent Transmittance: 28% – 90%
Absorbance: 0.050 – 0.550
Beer's law curves start to lose their linearity at absorbance values above 0.550 (percent transmittance values less than 28%). In case a solution that transmits such a low level of light, dilute the solution so that it falls within this range.

WAVELENGTH RANGES

One can select three LED light colors with the Vernier Software Colorimeter Blue (470 nm or 4700 A), green (565 nm or 5650 A) and red (635 nm or 6350 A). One can select one of these three nearly mono-chromatic colors using the wavelength selection knob on the top left of the colorimeter. There are several ways one can decide which of the three wavelengths to use. Look at the color of the solution. Remember that the color of a solution is the color of light which passes through it. One would probably want to use a different color of light that will be absorbed, rather than transmitted. For example, with a blue $CuSO_4$ solution, one should use the red LED (635nm)

CALIBRATION OF THE COLORIMETER

The calibration process involves two steps. In the first step, close the lid of the Colorimeter, set the knob on the Colorimeter to 0%T, and allows the reading on the CBL to stabilize. Press the button on the CBL and enter 0. Now set the knob on the Colorimeter to a wavelength and insert a blank cuvette. Allow the reading on the CBL to stailize and, again, press the button on the CBL. Enter 100. When calibration is complete, the CBL will be set up with the conversion equation, and you can now perform an experiment.

USING CUVETTES WITH THE COLORIMETER

The Colorimeter is designed to use polystyrene cuvettes. The cuvettes have a volume of approximately 4 mL. The cuvette slot of the colorimeter is designed to give a snug fit to the cuvette and ensure that it is always in precisely the same position between the LED light source and photodiode. Two opposite sides of the cuvette are ribbed and are not intended to transmit the light from the LED. The two smooth surfaces are intended to transmit light. It is important to position the cuvette correctly in the colorimeter. This should be done with the ribbed edges facing away from and toward you, and the smooth edges facing left and right. The light travels from left to right from the LED through the cuvette to the photo-diode. To avoid this, you should use a water-proof marker to make a reference mark on the right side of the top edge of the cuvette. This reference mark should be aligned with the white reference mark on the top right side of the colorimeter each time they insert a cuvette.

It is very important that solutions be added to a cuvette to the proper depth. A "safe level" is between 2.2 and 3.5 mL of solution. When the cuvette is filled to the brim, its total volume is about 4.1 mL. Since the inside diameter of the cuvette is about 1.0 cm x 1.0 cm, this safe level can also be measure on the outside of the cuvette as 2.2-3.5 cm from the inside bottom of the cuvette. These levels are shown in Figure.

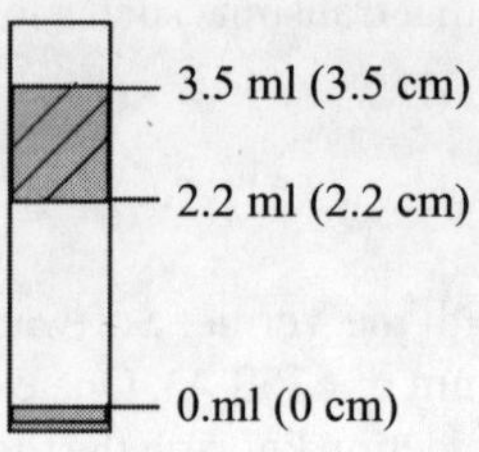

Fig. 37   A Colorimeter Cuvette

PARTS OF A COLORIMETER

A colorimeter consists of mainly six parts.

1. A light source
2. A condensing lens to render the light rays parallel
3. A filter to generate monochromatic light
4. A sample holder
5. A photocell to convert light energy (photons) into electrical energy.
6. A galvanometer to measure the electrical energy (current) thus generated.

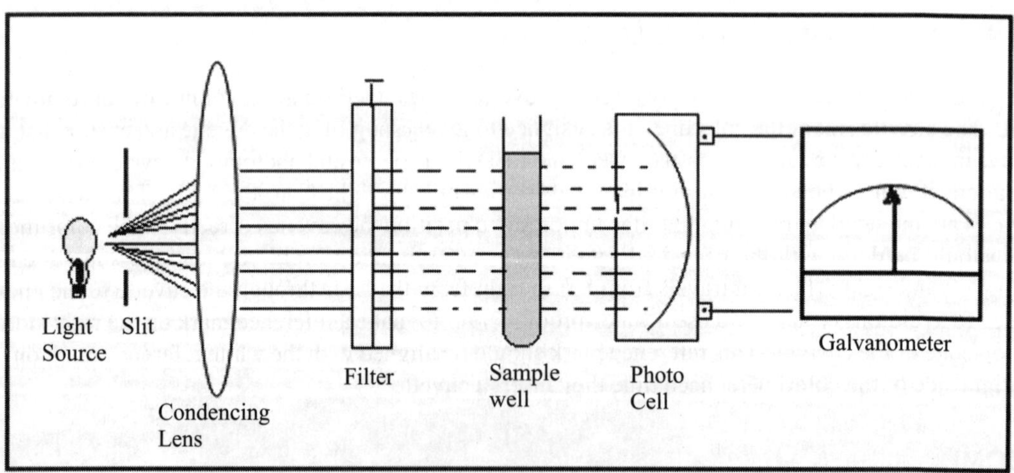

Fig. 38   Parts of a colorimeter

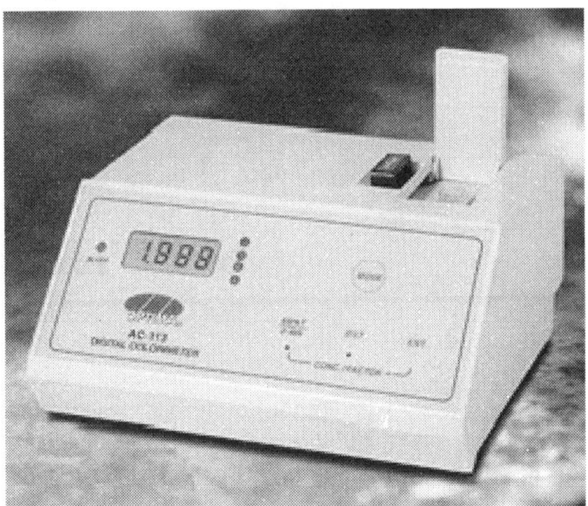

Fig. 39 A schematic diagram of colorimeter

The light from a tungsten lamp (composed of wavelengths between 400 and 900 nm) passes through a slit, a condensing lens, a filter and finally emerges as parallel beam of monochromatic light. The monochromatic light passes through the sample solution and the transmitted light falls in the photocell. The photocell converts the transmitted light energy into electrical energy, which is amplified and measured by the galvanometer. The galvanometer is calibrated to read the Absorbency/Transmittance directly.

COMPLEMENTARY COLOURS

When a coloured solution absorbs the light from a particular colour of the filter to a maximum extent, then that colour is said to be complementary to the colour of the solution (see Table)

Table: Complementary colour for some coloured solution

Colour of the Solution	Range of Wavelength (nm)	Complementary or Subtraction colour
Violet	400 – 465	Greenish Yellow
Blue	465 – 482	Yellow
Green	498 – 530	Red Purple (magenta)
Yellow	576 – 580	Blue
Orange	587 – 610	Greenish blue
Red	617 – 660	Bluish green
Purple Red	670 – 720	Green

COMPLEMENTARY WAVELENGTH

A complementary wavelength (λ_{max}) of the solution is the wavelength of a coloured solution that absorbs light maximally at a particular wavelength. Complementary wavelength is used to estimate the concentration of the substances because at that colour or wavelength only, the given substance absorbs maximally and contributes to high sensitivity to the estimations and also less interference from other substances

39

SPECTROPHOTOMETER

Spectrophotometer is also a type of photometer, a more refined instrument and it gives a far better precision and resolution than a colorimeter. It finds a variety of applications in biological research. A spectrophotometer essentially measures light absorption as a function of wavelength. Light is a form of energy and is propagated in the form of waves. Many characteristics of light such as reflection, fraction, diffraction, interference and its traveling in space, can be explained by assuming light as a wave. These light waves are considered to be a packet of energy or photons. Max-planck discovered that the light energy is directly proportional to the velocity of light and inversely proportional to its wavelength (λ)

i.e., $$LE \, \alpha \, \frac{c}{\lambda} \text{ or (where LE, is the light energy)}$$

i.e. $$LE = \frac{hc}{\lambda} \; ; \text{ since } = \frac{c}{\lambda}$$

therefore, $$LE = hv$$

Where 'c' is the velocity of light in vacuum (3×10^8 m/sec.), 'h' is the Planck constant (6.627×10^{-27} erg. sec.) and 'μ' is the frequency (i.e.) number of oscillations per sec (Hertz).

[One mole of radiation is equal to 6.023×10^{23} photons (Avagadro number of photons) and is equal to one Einstein].

APPLICATIONS

1. It is used to estimate the concentration of both coloured as well as colourless solutions, which could absorb light.
2. Because of its higher sensitivity it is used to estimate extremely small quantities of substances in a matter of a few minutes.
3. It usually does not degrade or modify the materials studied (unless a photochemical reaction occurs) and hence the materials can be recovered and reused.
4. It is also used to find out absorption maxima of compounds in a wide range of wavelengths.
5. It offers selectivity in that each component in a solution or reaction mixture can be singled out and estimated.
6. It also enables one to follow details of fast reactions and fast enzymes kinetics.
7. It is also used to measure the growth of bacteria and yeasts and to determine the number of cell in a culture.
8. Small volumes (as small as 0.3 ml) can be used for estimation of precious samples.

PRINCIPLE OF SPECTROPHOTOMETER

A spectrophotometer consists of two instruments, namely a *spectrometer* for producing light of any selected color (wavelength), and a *photometer* for measuring the intensity of light. The instruments are arranged so that liquid in a cuvette can be placed between the spectrometer beam and the photometer. The amount of light passing through the tube is measured by the photometer. The photometer delivers a voltage signal to a display device, normally a galvanometer. The signal changes as the amount of light absorbed by the liquid changes.

If development of color is linked to the concentration of a substance in solution then that concentration can be measured by determining the extent of absorption of light at the appropriate wavelength. For example hemoglobin appears red because the hemoglobin absorbs blue and green light rays much more effectively than red. The degree of absorbance of blue or green light is proportional to the concentration of hemoglobin.

When monochromatic light (light of a specific wavelength) passes through a solution there is usually a quantitative relationship (Beer's law) between the solute concentration and the intensity of the transmitted light, that is,

$$I = I_0 \times 10^{-Rc}$$

where I sub 0 is the intensity of transmitted light using the pure solvent, I is the intensity of the transmitted light when the colored compound is added, c is concentration of the colored compound, l is the distance the light passes through the solution, and k is a constant. If the light path l is a constant, as is the case with a spectrophotometer, Beer's law may be written,

$$I \div I_0 = 10^{-Rc} = T$$

Where k is a new constant and T is the transmittance of the solution.

There is a logarithmic relationship between transmittance and the concentration of the colored compound. Thus,

$$-\log T = \log 1/T = kc = \text{optical density (O.D.)}$$

The O.D. is directly proportional to the concentration of the colored compound. Most spectrophotometers have a scale that reads both in O.D. (absorbance) units, which is a logarithmic scale, and in % transmittance, which is an arithmetic scale. As suggested by the above relationships, the absorbance scale is the most useful for colorimetric assays.

$$1 \text{ MHz} = 1000 \text{ KHz or 1 million Hertz (cycles per second). } (10^6 \text{ nm} = 1\text{cm})$$

In double beam spectrophotometers, if you place a half-silvered mirror on its path, the monochromatic light coming out from the lens is split into two halves. Now 50% of the light passes directly through the mirror and falls on the reference cuvette and 50% of the light is reflected onto a second silvered mirror and then allowed to fall on the sample cuvette. At any given time the intensities of the transmitted lights from the reference and sample cuvette. At any given time the intensities of the transmitted lights from the reference and sample cuvettes are measured, amplified, the difference in intensities computed and sent to the Readout.

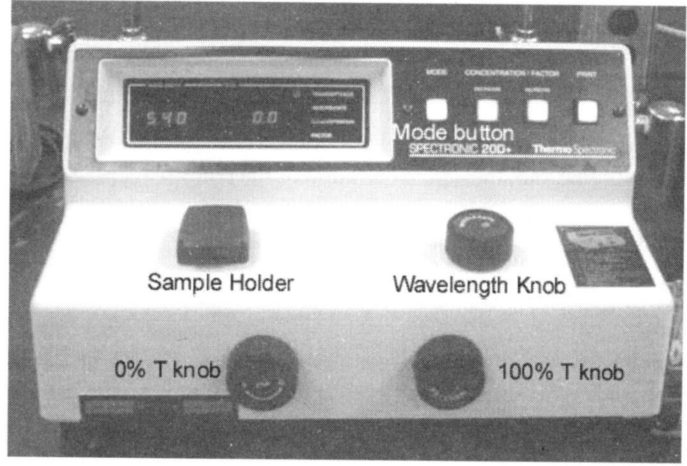

Fig. 40 Parts of a spectrophotometer

PREPARATION OF A STANDARD CURVE

Let us set up a hypothetical assay to measure substance X. When X is mixed with assay reagent a complex is formed that absorbs light at wavelength 400 nm. Take 2 ml volume in each cuvette. A cuvette is a transparent vessel to be placed in a light path for measurement of absorbance. To get the right proportion of assay reagent to sample, make the sample volume 0.5 ml and add 1.5 ml color reagent to each tube. Set up in this manner the assay can detect amounts of X of as little as 10 micrograms (μg) to as much as 2 milligrams (mg).

REFERENCE

To calibrate the spectrophotometer a reference tube is required that is identical in every respect to the standards and samples, except that it does not contain any substance X. With the light path blocked the spectrophotometer will be set to read infinite absorbance (no transmittance of light at all). With the reference tube in the light path set the spectrophotometer to read zero absorbance. That way, a sample containing X will give absorbance within that range. The reference tube is used to give us the maximum dynamic range. The reference will contain 0.5 ml sample buffer and 1.5 ml color reagent.

STANDARDS

To achieve the best accuracy a standard curve is to be set up with a logarithmic progression of standards. We need standards from 0.01 mg to 2 mg. Make the following dilutions 0.01, 0.02, 0.05, 0.1, 0.2, 0.5, 1, and 2 mg. The standards are prepared from a concentrated stock solution of the substance. The following table presents the calculations.

amount of substance X (mg)	volume of stock solution (μl)	volume of buffer (μl)
0 (reference)	0	500
0.01	2	498
0.02	4	496
0.05	10	490
0.1	20	480
0.2	40	460
0.5	100	400
1	200	300
1.5	300	200
2	400	100

The volume of buffer is not as critical as the volume of stock solution. Errors in pipetting buffer affect total volume and, thus, concentration of the color reagent. Errors of less than 1% will not have a significant effect on the results. It may be necessary to conduct a serial dilution to get, say, 2 or 4 μl stock solution into an assay tube.

SAMPLE PREPARATION

It helps to have a reasonable estimate of ranges of concentrations of sample that one can expect. Even with such an estimate it is good to prepare samples with a range of dilutions, in case a sample is so concentrated that its absorbance readings are out of range. For this assay if 500 μl sample is used in an assay tube (the maximum volume), its concentration should be less than 4 mg/ml to give a readable absorbance. Knowing nothing about the concentration of a particular sample, we would load one tube with 500 µl to cover that range. Since the assay spans a wide range of concentrations, we can use 50 µl in a second assay tube. Now the sample can be as concentrated as 40 mg/ml and we will still have 4 mg or less in the assay tube, giving a readable result. To cover all of the bases we can assay a third tube with just 5 µl sample.

RUN THE ASSAY

When all of the standard and unknowns are ready we will have:

- 1 reference tube
- some number of standards that span the range of the assay
- three assay tubes per sample representing a series of dilutions

For colour development one has to add a color reagent and letting the samples sit for a few minutes. The instrument should be calibrated, and then absorbances should be taken for each tube in order. A standard curve is obtained by plotting absorbance versus amount of substance X. If the relationship is clearly linear, a standard curve isn't even necessary. A curve should be constructed the first time an assay is used, to check for accuracy and linearity.

EXAMPLE OF A STANDARD CURVE

Here is what the plot might look like in a graph paper.

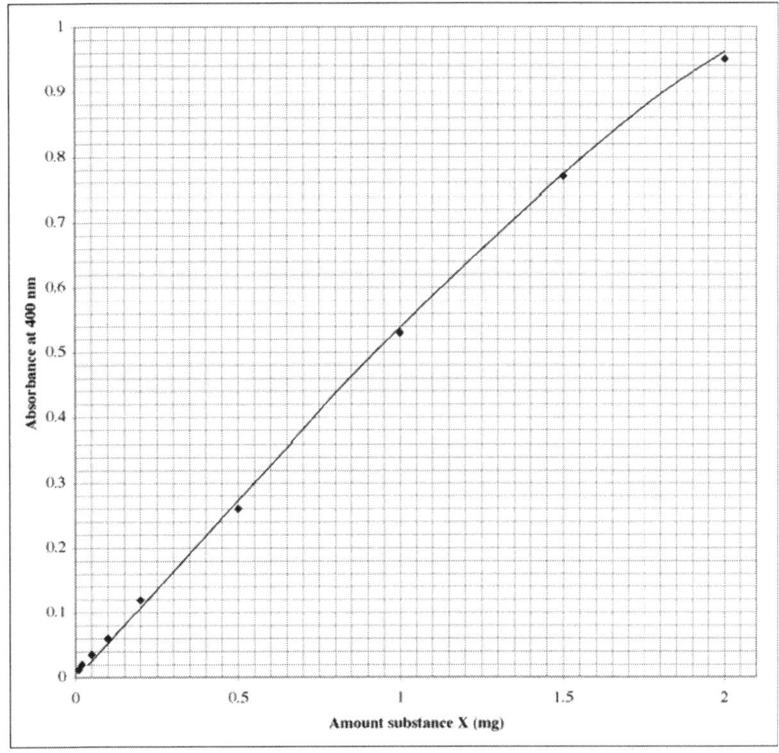

Determination of concentration of a sample

Let us suppose that one has prepared three assay tubes for sample 1, containing 500 µl, 50 µl, and 5 µl samples, respectively and they gave absorbance readings of 0.86, 0.12, and 0.01, respectively. Absorbance of 0.86 corresponds with 1.7 mg substance X. The volume was 500 µl (0.5 ml), so we get a concentration of 3.4 mg/ml. Checking the other readable tube, absorbance of 0.12 indicates that the tube contained 0.20 mg substance X. The volume was 50 µl (0.50 ml). The concentration should be 0.20 mg/ 0.050 ml = 4.0 mg/ml. Which result do we use, or do we take an average?

It has been found that using the one absorbance reading that falls closest to the middle of the sensitive range gives the most accurate results. In the example above the center is an asorbance of 0.5, corresponding to 0.1 mg of substance. The absorbance scale is logarithmic.

USING A SPECTROPHOTOMETER

- The instrument must have been warm for at least 15 min. prior to use. The power switch doubles as the zeroing control.
- Use the wavelength knob to set the desired wavelength. Extreme wavelengths, in the ultraviolet or infrared ranges, require special filters, light sources, and/or sample holders (cuvettes).
- With the sample cover closed, use the zero control to adjust the meter needle to "0" on the % transmittance scale (with no sample in the instrument the light path is blocked, so the photometer reads no light at all).
- Wipe the tube containing the reference solution with a lab wipe and place it into the sample holder. Close the cover and use the light control knob to set the meter needle to "0" on the absorbance scale.
- Remove the reference tube, wipe off the first sample or standard tube, insert it and close the cover. Read the absorbance.
- Remove the sample tube, readjust the zero, and recalibrate if necessary before checking the next sample.

A proper reference solution contains color reagent plus sample buffer. The difference between the reference and a sample is that the concentration of the assayable substance in the reference solution is zero. The reference tube transmits as much light as is possible with the assay solution one is using. A sample tube with any concentration of the assayable substance absorbs more light than the reference, transmitting less light to the photometer. In order to obtain the best readability and accuracy, the scale is set to read zero absorbance (100% transmission) with the reference in place. If you use a water blank as a reference, you might find that the assay solution alone absorbs so much light relative to distilled water that the usable scale is compressed, and the accuracy is very poor.

40

ESTIMATION OF PROTEIN

THEORY

Proteins one of the important macromolecules found in all living cells, are nothing but polymers of different amino acids (the word 'Protein' is derived form the Greek word proteinous which means "of primary importance", coined by JJ Berzelius in 1838). These are about twenty different amino acids, which make these proteins. These amino acids are organic, amphoteric molecules having an amino group at one end and a carboxylic group at the other end. In proteins the amino acids are connected by peptide bonds as shown in figure below.

$$
\begin{array}{ccccc}
R_1 & R_2 & R_3 & R_4 & R_5 \\
| & | & | & | & | \\
\end{array}
$$

$$-CO-NH-CH-CO-NH-CH-CO-NH-CH-CO-NH-CH-CO-NH-CH-CO-NH$$

Peptide bond in protein chain

Where $R_1 - R_5$ represents side chains in different amino acids

Various methods such as Biuret, UV, Lowry, Dye binding and Fluorescamine are available to determine the concentration of proteins in biological samples. Each of these methods vary in their sensitivity and hence their applicability.

Calibration:

The test tubes to be used for protein estimation should be cleaned thoroughly since even a fingerprint may contain up to 0.1 mg protein. The test tube are soaked in 5 M HCl or HNO_3 overnight, washed thoroughly and dried in an oven. Weigh ~ 10 mg bovine serum albumin (BSA) and dissolve in 10 ml of double distilled water gently by inverting the tube. Once completely dissolved, measure the absorption of the BSA stock solution at 280-mm. Exact concentration of BSA (mg/ml) is arrived at by dividing the A_{280} by 0.66, (specific extinction coefficient of Pure BSA is 0.66).

1. Biuret Method

The name of the test comes from the compound known as biuret, which gives a typical positive reaction for the test.

$$CONH_2$$

$$NH \qquad \text{(Structure of biuret)}$$

$$CONH_2$$

This method is absolutely specific for peptide bonds and hence assays the number of peptide bonds in proteins.

PRINCIPLE

Cu^{2+} in alkaline solution complexes with nitrogen atoms of the peptide bonds in proteins as shown in Figure below. This complex forms a purple colour at high temperature. (Cu^2 is maintained in alkaline solution as the tartarate complex). The colour is measured at 535-560 nm of near UV region at 310 nm.

$$
\begin{array}{cccc}
R_1 & R_2 & R_3 & R_4 \\
| & | & | & | \\
-CO-NH-CH-CO-NH-CH-CO-NH-CH-CO-NH-CH-CO-
\end{array}
$$

$$Cu^{++} \qquad\qquad\qquad\qquad Cu^{++}$$

$$
\begin{array}{cccc}
-CO-NH-CH-CO-NH-CH-CO-NH-CH-CO-NH-CH-CO- \\
| & | & | & | \\
R_1 & R_2 & R_3 & R_4
\end{array}
$$

Copper atoms of the reagents forming coordinate complexes with the N-atoms of the peptide bonds

REAGENTS

Dissolve 1.5 g $CuSO_4.5H_2O$ and 6.0 g Na – K tartarate.$4H_2O$ (Rochelle salt) in 500 ml of boiled and cooled water (The water is boiled and cooled to remove the dissolved oxygen which otherwise might convert Cu^{2+} to Cu^+, a reddish orange precipitate). This reagent is stable for months at room temperature. This reagent must be discarded if reddish orange precipitate (Cu^+) forms.

PROCEDURE

Pipette 50 ìl to 500 ìl of protein solution (from a bovine serum albumin stock of 10 mg/ml) and make up the volume to 0.5 ml with distilled water. Add 5.0 ml colour reagent and place the tubes at 80° C water bath for 5 minutes. Cool rapidly by placing them in a tray containing tap water. Stand at room temperature for 10 minutes. Read the absorbency at 535 nm. 1 mg protein measures an Absorbency of about 0.1 at 535 nm. The colour is stable for several days if the solution is kept in dark. As the method is not sensitive, the method is not used routinely in labs.

Sensitivity: 1 – 5 mg.

2. UV Method

The method is based on the principle that the aromatic amino acids in proteins particularly tryptophan and tyrosine absorb light strongly at 280 nm.

Amino acid	λ_{max} (nm)	$\Sigma_{-max} \times 10^{-3}$ at pH 7.0
Trp	280	5.6
Tyr	274	1.4
Phe	257	0.2

If the protein solution is turbid, it should be clarified by configuration before estimation. Measure the absorbency in a buffered solution (pH 6 – 9) using the same buffer as blank. The method is not as specific as colorimetric methods, but it is nondestructive and the sample can be recovered and reused. Protein solutions at 1 mg/ml give an absorption of ~ 1 at 280 nm and ~ 15 at 215 nm. For a pure protein, the 280/260 = 1.8 and for pure nucleic acid the 280/260 = 0.5. If nucleic acid or nucleotides are present in the protein samples, the absorption at 280 nm and 260 nm are measured. Then the following equation is used to calculate the protein content.

$$\text{Protein (mg/ml)} = 1.55 \times A_{280} - 0.76 \times A_{260}$$

All proteins absorb strongly below 230 nm due to peptide bonds. Therefore, the 0.1% values are essentially the same for all proteins. For example, the following are the absorption values of a 0.1% protein solution with 1 cm light path at different wavelengths.

$$A_{210} = \sim 20,\ A_{215} = \sim 15,\ A_{225} = \sim 14 \text{ and } A_{280} = \sim 1.$$

Protein concentrations in the region of 10 -100 ìg/ml can be determined from the difference in absorbency at 215 and 225 nm.

$$\text{Protein (μg/ml)} = 144\ (A_{215} - A_{225})$$

Sensitivity: 200 – 2000 μg and 10 -100 μg.

Interfering agents: 0.1 M NaOH interferes at 215 nm but NaOH concentrations up to 5 mM cause no problem.

3. Lowry's Method

This is the most widely used method in all laboratories. In this method the colour development relies

(1) On the formation of a copper – protein complex as in biuret reaction.

(2) On the reduction of phosphomolybdate and phosphotungstate anions present in Folin – Ciocalteu phenol reagent by the tyrosine and tryptophan residues of the proteins, to heteropolymolybdenum blue and tungsten blue respectively, which gives a blue coloured complex with a λ_{max} of 750 nm.

(3) Cu^{2+} also acts as a catalyst in the reduction reaction.

The intensity of the colour mainly depends on the aromatic amino acid content of a protein. Approximately 75% reduction occurs due to the copper – protein complex with tyrosine and to lesser extent (25%) with tryphtopan residue.

REAGENTS

Reagent – A

2% Na_2CO_3 in 0.1 N NaOH.

Dissolve 4 g sodium hydroxide in 800 ml distilled water and then dissolve 20g sodium carbonate (anhydrous). Make up the volume to 1000 ml with distilled water and store at room temperature.

Reagent – B

2% $CuSO_4$. 5 H_2O
  Dissolve 2 g copper sulphate in 100 ml distilled water and store at room temperature.

Reagent – C

2% Sodium Potassium tartarate or 2% trisodium citrate.
  Dissolve 2 g of sodium potassium tartarate or trisodium citrate in 100 ml distilled water. Store at room temperature.

Reagent – D (Prepare fresh every day)

Mix 0.5 ml of Reagent B and 0.5 ml of Reagent C, and then add 99 ml Reagent A. If the Reagent D becomes turbid, do not use it. (Mix the reagent in the above order)

Reagent – E

Commercial Folin – Ciocalteau reagent (AR) is diluted with water to 1 N (1 N solution can be stored at 4° C for a week). Discard this reagent if the absorption of the stock solution at 660 nm is greater than 0.2.
Procedure: (For spectrophotometer estimation)

(1) Prepare 10 tubes containing increasing amounts of BSA (0.5 mg/ml) (0,5,10,20,30,40,50,60,75 and 100ìg) in a total volume of 0.2 ml (see Table 6.1)
(2) Prepare three tubes with increasing volumes of unknown sample in a total volume of 0.2 ml.
(3) Add 1.0 ml of reagent D (B+C+A) to each tube, vortex and let stand for 10 min at room temperature.
(4) Add 0.3 ml of reagent E (Folin's phenol reagent) and vortex the tube **immediately*** and let stand for 60 minutes RT.
(5) Read at 660 nm or 750 nm and plot the values A_{660} vs Concentration of BSA.

Ten µg protein gives the absorption of 0.1 at 750 nm. The linearity of the assay can be extended over the range of 30 – 200 µg by measuring at A_{550} rather than at A_{750}.

*Folin's phenol reagent is acidic. The half-life of the reagent in alkaline copper reagent is only about 8 seconds. Hence each tube is vortexed immediately on the addition of the reagent.

Table – Outline for Lowry's procedure (spectrophotometer)

Tube No. –	0	1	2	3	4	5	6	7	8	9	10
BSA (µl)-	0	5	10	20	30	40	50	60	75	100	150
H_2O(µl)-	200	195	190	180	170	160	150	140	125	100	50

Reagent D _____________________1.0 ml___________________________________

Vortex and incubate for 10 minutes at room temperature.

Reagent E _____________________0.3 ml___________________________________

Vortex the tubes immediately on each addition and incubate for 60 minutes at room temperature. Read absorption at 660 nm.

Sensitivity: 5-50 µg.

Interfering agents: Tris buffer, ammonium sulphate, sucrose and sulphydryl compounds (â-mercaptoethanol, dithiothreitol), glycine, urea, etc., are interfering agents. It is necessary to include proper blank containing the same amount of chemical present in the protein solution.

For colorimeteric estimations, make up the volume of the protein solutions to 1 ml with water. Add 4 ml alkaline copper reagent is and mix. After 10 minutes at RT, add 0.5 ml 1N Folin's phenol reagent and vortex **immediately.** The colour is measured at 660 nm after 1 hr at RT.

PREPARATION OF FOLIN – CIOCALTEAU REAGENT

Into a two – litre round bottomed flask, add 500 ml of distilled water and dissolve 100 g of sodium tungstate and 25 g of sodium molybdate. Add slowly 50 ml of 88% phosphoric acid and 100 ml of concentrated hydrochloric acid by shaking the flask gently during the addition of the acids. Reflux the mixture gently for about 12 hours with a condenser. Cool the flask to room temperature and add 150 g of lithium sulphate, 50 ml distilled water and few drops of bromine. Mix and boil the flask for 10 minutes without the condenser, to remove excess bromine. Cool the flask to room temperature and make up the volume to one litre with distilled water. Filter, if necessary, using a Whatman 1 filter paper on a glass funnel. The filtrate should not have any greenish tinge. If there is, it has to be boiled again with a few drops of bromine. This is the stock reagent and is stored in brown bottles at 4°C.

TITRATION

The normality of the stock solution will be usually between 1.8 N to 2.2 N. In order to find out the exact normality of the solution, it is titrated against 1N NaOH* using phenolphthalein as indicator. The required volume of stock solution can be diluted to 1 N with distilled water just before protein estimation. This reagent is also available commercially. *titrate the 1 N NaOH with a primary standard.

4. BRADFORD'S METHOD

This has also become a popular method for estimation of proteins because the assay is simple, quick and inexpensive. This method is based on the principal that proteins bind Coomassie brilliant blue G-250 in acid solution and form a complex whose extinction coefficient (λ_{max} = 595 nm) is much greater than the free dye (λ_{max} = 465 nm) itself.

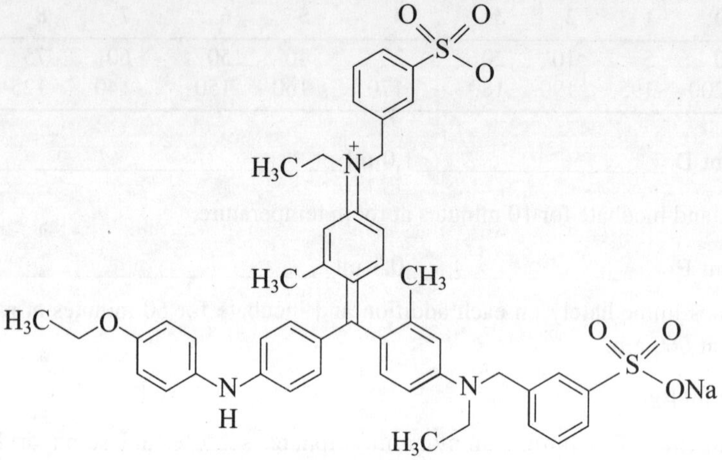

Commasie brilliant blue

    The Coomassie brilliant blue G-250 or the Serva blue-G appears as a ple – orange red in protonated form (i.e.), in acid solution. The dye binds strongly to positively charged groups of proteins (especially to Arg residues and to a lesser extent to Lys, His and aromatic amino acids) and also to hydrophobic regions in proteins. As a result, a blue colour is formed with a λ_{max} at 595 nm. (On binding to proteins the λ_{max} is shifted from 465 to 595 nm).

PREPARATION OF BRADFORD'S REAGENT: (2 X)

(It is convenient to prepare a 2 x stock and dilute to read an absorbency of ~1.1)

Dissolve 100 mg of G-250 in 100 ml of absolute/distilled ethanol or absolute methanol in on a shaker for ~ 60 min. Add 100 ml 88% O-phosphoric acid, mix well and make the volume to 500 ml with distilled water. Filter through Whatman No. 1 paper. Dilute 1:1 with water and check the A_{550} against water blank. The A_{550} should be ~ 1.1. If not adjust the dilution.

PROCEDURE

A. For spectrophotometric estimation

(1) Pipette out increasing amounts (10-100 µl) of BSA from a 0.2 mg/ml stock solution into clean dry test tubes (2 µg to 20 ìg range).
(2) Make up the volume to µl with distilled water. Keep a blank
(3) Add 1.5 ml of the dye solution to each tube and mix gently.

(4) Leave the tubes for 5 minutes at room temperature.
(5) Read the colour at 595 nm.

B. For colorimetric estimation:

(1) Pipette out 10 – 100 µl of the BSA stock solution (0.2 mg/ml)
(2) Make up the volume to 0.5 ml with distilled water. Keep a blank.
(3) Add 4.5 ml of the dye reagent and mix gently
(4) Read the absorbency at 595 nm after 5 min at RT.

Note

Five ìg proteins read an absorbency of ~ 0.15 at 595 nm. The colour is stable for 30 min; after that a precipitation of the dye – binding complex occurs. The cuvettes and the test tubes should be washed with concentrated hydrochloric acid or crude alcohol before washing with detergents.

Sensitivity: 1 – 20 ìg.

INTERFERING AGENTS

This method is less prone to interference with other agents. But anionic detergents like SDS and deoxycholate interferes. But Triton X – 100 does not interfere. (If the protein is insoluble it can be solubilized using Triton X – 100 which does not interfere.) Strong alkaline solutions cause false positive results.

5. FLUORESCAMINE METHOD

This is the most sensitive method to estimate the concentration of proteins. Fluorescamine is a heterocyclic dione. It reacts with primary amines to form fluorescent product. In proteins or peptides it reacts with the free N- terminal amino groups. Hence the reaction is carried out at pH 8.5 to 10.0 to suppress ionization of the amino groups of side chains.

Prepare stock solutions free from traces of water because in presence of water fluorescamine hydrolyses to give non-fluorescent products. Half of the molecules are hydrolyzed in 1-10 sec.

Reagents

(a) Borate buffer, pH 8.5: (dissolve 3 g boric acid and 4.8 g sodium tetraborate in 200 ml of water.
(b) Fluorescamine solution, 0.03%: (dissolve 7.5 mg fluorescamine in 25 ml Analar grade acetone). This solution is stable at room temperature.

Procedure

1. Prepare tubes containing increasing concentrations of BSA (0.1 to 2 ìg) in clean dry tubes.
2. Make up the final volume to 0.1 ml with distilled water.
3. Add 0.9 ml borate buffer to each tube and vortex.
4. While vortexing add 0.1 ml fluorescamine solution.
5. Read the fluorescamine after 2 minutes using $\lambda_{ext} = 390$ nm and $\lambda_{max} = 465$ nm.

Sensitivity: 0.1 µg – 2 µg

Note

This method is particularly useful to detect small peptides with molecular weights less than 10,000.

Interfering agents

This method is less prone to interfering agents except buffers made up of amines such as Tris or ammonium bicarbonate.

Important point to be noted

Other compounds of biological origin like carbohydrates, lipids, pigments and polyphenols are also known to interfere with the estimation of proteins. To extract carbohydrates and pigments from protein samples, add equal volume of cold 10% trichloroacetic acid (TCA), leave it on ice for 10 minutes, centrifuge and dissolve the pellet in 0.1N NaOH. This solution is used for the estimation of proteins.

If there is interference with lipids, the sample is extracted once with a mixture containing ethanol and ether (2:1 v/v), centrifuged and the upper organic phase is discarded. The material is extracted once with ether, centrifuged and upper ether phase is discarded. The material is air dried to remove excess solvent. The air-dried material can now be extracted with equal volume of cold 10% TCA solution to precipitate out the proteins, as described above.

Plant tissue extracts contain considerable amount of polyphenols, which also get precipitated by TCA and interfere with protein estimation. Therefore, when plant tissues are used, it is a general practice to include polyvinyl pyrrolidone (PVP) to a final concentration of 1% in the extraction buffer. PVP forms a complex with polyphenols and the precipitate thus formed can be removed by centrifugation.

Cold TCA at a final concentration of 5% precipitates the proteins. But TCA does not precipitate the proteins if the protein concentration is less than 100 ìg/ml. In such cases the proteins can be precipitated by TCA in presence of a carrier molecule such as sodium deoxycholate.

To 1 ml dilute protein solution, add 0.15 ml of 0.1% sodium deoxycholate in distilled water. Allow the solution to stand for 10 minutes at room temperature. Add 0.2 ml of 50% (w/v) cold TCA. Allow it to stand for 15 min on ice. Pellet the protein by centrifuging in an Eppendorf centrifuge for 10 min. Discard the supernatant and leave the tubes inverted on tissue paper until all the liquid is drained. Remove the droplets with tissue paper. Dissolve the pellet in ~ 20 ìl of 10 mM NaOH.

Note

Deoxycholate acts as a carrier molecule when the protein concentration is very low. Protein concentrations as low as 5 μg/ml can be concentrated and estimated using the deoxycholate method.

41

ESTIMATION OF AMINO ACID

Amino acids are routinely estimated by ninhydrin method. Ninhydrin is a powerful oxidizing agent. It reacts with all á – amino acids betwwn pH 4-8 and produces a purple coloured complex. This reaction is also given by primary amines, imino acids and ammonia but without the liberation of CO_2. The amino acids praline and hydroxyl-proline also react with ninhydrin but they produce yellow coloured compound. The purple colour is measured at 550 nm whereas the yellow colour is measured at 440 nm. The ninhydrin reaction is very sensitive and ideal for the detection of ìg quantities of amino acids in chromatograms (Paper and TLC) and also in column fractions.

Ninhydrin + CH3—CH(NH2)—COOH $\longrightarrow$ Reduced Ninhudrinv + CH_3 + RCHO + CO_2

Ninhydrin   Amino Acid   Reduced Ninhudrinv

Ninhydrin   Reduced Ninhydrin

C=N + $3H_2O$

Coloured Complex (Ruhemann's Purple)

REAGENTS

1. 0.5 M acetate buffer, pH 5.5
2. Ninhydrin reagent: Dissolve 2.0 – g ninhydrin in 30 ml of acetone and add 20 ml of 0.5 M acetate buffer, pH 5.5. Prepare fresh every time. Store in brown bottle.
3. Stock solution of amino acids: 1 mg/ml in dH_2O.

PROCEDURE

1. Pipette out 10 to 50 ìl of the amino acid stock solution.
2. Make the volume to 4.0 ml with distilled water.
3. Add 1.0 ml freshly prepared ninhydrin reagent.
4. Cover the tubes with aluminum foil and place them in a boiling water bath for 10 minutes.
5. Cool to room temperature in a trough containing tap water and 1 ml 50% ethanol to each tube.
6. Allow it to stand for 5 minutes and read the colour at 550 nm.

Note:

The colour fades with time. Lower concentrations and dilute solutions do not produce any colour.

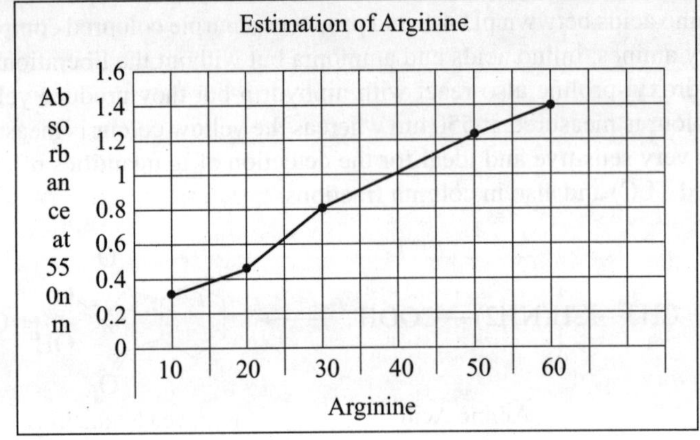

42

ESTIMATION OF CARBOHYDRATE

THEORY

Carbohydrates are one of the important macromolecules found in all living cells. They are the major constituents of all cell walls. They involve in cell wall formation and cell-cell recognition. They also serve as storage polysaccharides and antigenic determinants. Moreover, most of the proteins in eukaryotic organisms are glycoproteins and carry carbohydrates on their peptide backbone.

The carbohydrates are polyhydroxy aldehydes or ketones or substances, which yield such compounds upon hydrolysis. Since the empirical formula $C(H_2O)_n$ of these compounds suggests that these are hydrates of carbon, they are called carbohydrates. (this definition is no more valid now, as compounds containing nitrogen, phosphorous or sulphur also have the general properties of carbohydrates).

There are three major classes of carbohydrates viz., monosaccharides oligosaccharides and polysaccharides. (The word saccharide is derived fromGreek word, which means sugar). Monosaccharides consist of a single polyhydroxy aldehyde or ketone unit and cannot be hydrolyzed further into smaller units under mild conditions eg., trioses, tetroses, pentoses, hexoses, etc. The most abundant monosaccharides are glucose and fructose.

Oligosaccharides (Oligos (GK) = few) consist of a short chain of monosaccharides joined together by covalent bonds which are known as glycosidic bonds (just like amino acids are joined together by peptide bonds and nucleotides are joined together by glycosidic bonds) Oligosaccharides, in general, contain two to six molecules of monosaccharides e.g., raffinose, melezitose, stachyose, etc. Polysaccharides are very long polymers having hundreds or even thousands of monosaccharide units. The polysaccharides may be a linear polymer as in cellulose or a branched polymer as in starch or glycogen.

PROPERTIES OF SUGARS

Carbohydrates have a number of alcoholic groups, both primary ($- CH_2OH$) and secondary ($>CHOH$). This makes the molecule highly water-soluble. They also have a highly reactive aldehyde or keto group at one end of the molecule, which makes them powerful reducing agents. This reducing property of the sugars is useful in quantification of these sugars. Moreover, the presence os asymmetric carbon atoms (i.e.), the linking of the four valences of the carbon atom by four different groups, makes the sugars optically active. That is, they rotate the plane polarized light towards right (d – series) or towards left (l – series)

There are verities of methods available to estimate sugars. Most of these methods essentially make use of the reducing property of these sugars. In general, reducing sugars are estimated by Nelson – Somogyi's method (where Cu^{2+} is reduced by the sugar Cu^+) or dinitrosalicylic acid method (where the sugars are oxidized to their respective sugar acids). The monosaccharides can also be estimated specifically and more precisely using enzymes. The non-reducing sugars or total sugars are esti-mated by anthrone or phenol methods. The oligo- and polysaccharides are estimated, first by hydrolyz-ing them using enzymes or mineral acids to their respective monosaccharides and then estimated as mentioned above.

1. ESTIMATION OF SUGARS BY DINITROSALICYLIC ACID METHOD

PRINCIPLE

In alkaline solutions the reducing sugars from enediols. The enediols are readily oxidized to their respec-tive sugar acids by the oxidizing agent 3, 5 dintrosalicyclic acid (DNSA) as shown,

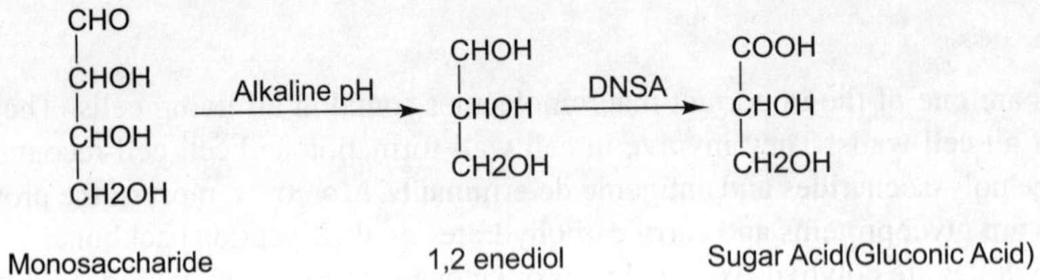

$$
\begin{array}{ccccc}
\text{CHO} & & \text{CHOH} & & \text{COOH} \\
| & & | & & | \\
\text{CHOH} & \xrightarrow{\text{Alkaline pH}} & \text{CHOH} & \xrightarrow{\text{DNSA}} & \text{CHOH} \\
| & & | & & | \\
\text{CHOH} & & \text{CH2OH} & & \text{CH2OH} \\
| & & & & \\
\text{CH2OH} & & & & \\
\end{array}
$$

Monosaccharide   1,2 enediol   Sugar Acid(Gluconic Acid)

REAGENTS

(a) Dinitrosalicylic acid reagent

Dissolve 1 g DNSA, 0.2 g phenol, 0.05 g sodium sulphite in about 80 ml dH_2O. Add 5 ml of 5N NaOH. Make up the volume to 100 ml.

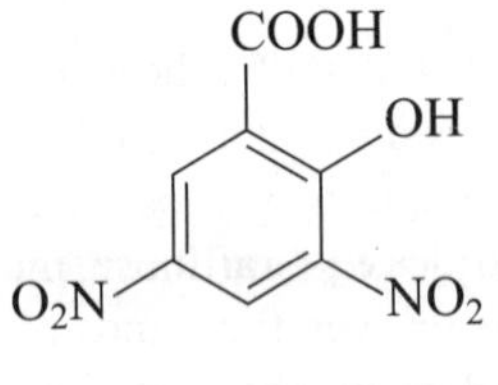

Dinitrosalisylic Acid

(b) Standard sugar solution:
Glucose or maltose is prepared at a concentration of 1/mg/ml in dH_2O.

PROCEDURE

Pipette out 0.0 ml (blank) to 3.0 ml of the sugar solution into clear test tubes. Make up the final volume to 3.0 ml with dH_2O in all the tubes. To each tube add 3.0 ml of dinitrosalicylic acid reagent and cover the with tubes with aluminum foil. Place the tubes in boiling water for 6 minutes and cool them in a trough containing tap water. The colour is measured at 575 nm (green filter) using the reagent blank. 0.1 mg glucose = ~ 0.1 A_{575}.

2. NELSON-SOMOGYI'S METHOD

PRINCIPLE

This is one of the widely used methods for the estimation reducing sugar. This method is based on the principle that reducing sugar, reduce copper sulphate (Cupric, Cu^{++}) in the reagent to red-orange cuprous (Cu^+) oxide (Cu_2O) precipitate. The cuprous oxide on reaction with arsenomolybdate reagent produces a green coloured complex, which is measured at 520 nm. (The λ_{max} is 660 nm, where the sensitivity is 4 times higher between sensitivity and the advantages gained by reducing to a minimum, the effect of variation due to reagents, reoxidation of cuprous oxide, etc). The colour is stable foe at least 18 hours.

$$CuSO_4.5H_2O + \text{Reducing Sugar} ? Cu_2O + \text{Sugar (Oxidized)}$$

$$Cu_2O + \text{Arsenomolybdate reagent} ? \text{Green coloured complex} + CO_2 ?$$

$$\lambda_{max} = 520 \text{ nm}$$

REAGENTS

1. Peparation of copper reagent A:

Solution I

Sodium carbonate (anhydrous)	– 24 g
Sodium potassium tartarate	– 12 g
Dissolve in ~ 40 ml distilled water	

Solution II

$CuSO_4 . H_2O$	– 4 g
Dissolve in ~ 40 ml distilled water	

Solution III

Add solution II to solution I slowly with stirring and then add 16 g $NaHCO_3$ and dissolve to make solution III.

Solution IV

Dissolve 180 g sodium sulphate in 500 ml of hot water and then boil the solution to expel air. (otherwise oxygen in the air will reduce the cupric ions to cuprous ions and give high background readings, during reducing sugar estimations). After cooling to room temperature, add the solution IV to the solution III and make the volume to 1 litre. After a week small amount of cuprous oxide that is settled at the bottom is removed by filtration using Whatman No. 1 paper and then used. The solution is stable at room temperature.

3. PREPARATION OF ARSENOMOLYBDATE COLOUR REAGENT

Dissolve 25 g ammonium molybdate in 450 ml distilled water (the solution will be turbid and will clear on addition of acid), add slowly 21 ml of conc. Sulphuric acid and mix with a glass rod. To the above solution add slowly with mixing, 3 g of sodium arsenate ($Na_2HASO_4.\ 7H_2O$) solution, already dissolved in about 25 ml water. Make the volume to 500 ml and place in an incubator at 37°C for 24 to 48 hours. If the reagent is needed quickly an alternative procedure is to incubate to 55°C for 30 minutes. Store in brown bottles at room temperature.

4. STANDARD SUGAR SOLUITIONS

Dissolve any reducing sugar at 1 mg/ml in distilled water.

PROCEDURE

Pipette out 0, 0.1, 0.2,.......1.0 ml of sugar solution in clean dr test tubes. Make up the final volume to 2.0 ml with distilled water. Add 2 ml of copper reagent to each tube. Place the tubes in a boiling water bath for 10 minutes and then cool in a trough containing tap water. Add 2 ml of arsenomolybdate colour reagent to each tube and vortex (colour develops very rapidly with evolution of CO_2). Read the colour at 520 nm using the blank. The colour is very stable.

43

ESTIMATION OF TOTAL SUGAR

1. ANTHRONE / PHENOL METHODS

This method can be used for estimation of reducing as well as non-reducing sugars (i.e.) total sugars in samples. In this method, the sugars in presence of conc.sulphuric acid gets dehydrate and produce furfural (from pentose) or 5-hydroxymethylfurfural (from hexoses), as shown below in Figure, which when reacted with anthrone or phenol produces a coloured compound with λ_{max} of 625 nm. Pentose, hexoses, heptoses and their derivates yield a coloured product in these reactions whereas trioses, tetroses and aminosugars do not yield any coloured product.

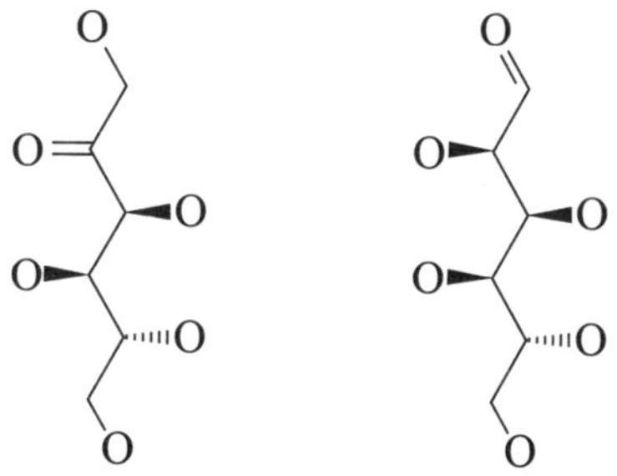

D-Fructose   D-Glucose

Note: Anthrone and phenol methods are simple, insensitive to interference and hence give a reliable index of total carbohydrates in samples.

REAGENTS

(a) Anthrone reagent: (0.2%) (Prepare fresh)
   Dissolve 200 mg anthrone in 5 ml distilled ethanol and make up to 100 ml with 75% sulphuric acid. For phenol reagent, dissolve 5 g phenol (AR) in 100 ml d H_2O.
(b) Standard sugar solution: (0.1%)
   Dissolve 100 mg of sugar (glucose, sucrose, methyl glucoside, etc.) in 100 ml distilled water. Dilute 1:10 for estimations.

PROCEDURE

Pipette out into a series of test tubes, increasing volumes of sugar solution from 0.0 ml – 1.0 ml and make up the volume to 0.1 ml with distilled water*. To each tube add 5.0 ml of cold anthrone reagent and vortex rapidly. Cover the tubes with aluminum foil and keep them in a boiling water bath for 10 minutes and cool to room temperature in a tray containing tap- water. Read the absorbency at 625 nm using the blank.

   *For phenol reagent mix 1 ml phenol reagent with the samples and then with 5 ml conc. Sulphuric acid. Place the tubes in a water bath at 37°C for 20 min., read the absorbency at 490 nm.

2. ESTIMATION OF SUGARS USING ENZYMES

PRINCIPLE

Unlike the chemical methods described earlier, enzymatic methods provide high degree of specificity in estimating, especially, blood glucose, as D-galactose, D-mannose and D-fructose do not react. Furthermore the enzymes is specific for â- D glucose as with á- D glucose, the reaction is ~ 1% only. The enzymatic procedure yields approximately the same results as the Nelson-Somogyi's method but the precision and specificity with the enzymatic procedure is far better.

$$B - D- \text{Glucose} + 2H_2O + O_2 \xrightarrow[\text{(oxidized)}]{\text{GOD}} \text{Gluconic acid} + 2\ H_2O_2$$

$$H_2O_2 + O\text{- dianisidine} \xrightarrow{\text{POD}} \text{Oxidized dianisidine} + H_2O$$

   The intensity of the brown colour is measured at 425 – 475 nm. Addition of HCl produces a yellow colour, which is measured at 420 nm. (Instead of O-dianisidine, O-toluidine can also be used).

REAGENTS

1. Glucose standard solution:

Dissolve 5 mg/ml in distilled water (store in a refrigerator).

2. O – Dianisidine dihydrochloride solution:

Dissolve 5 mg/ml in distilled water (the dye solution is stable for 3 months at 4°C).

3. Enzymes solution:

Mix 300 units of glucose oxidase and 100 purpurogallin units of peroxidase (4 mg of peroxidase) in 100 ml 0.1 M sodium phosphate buffer pH 7.0. (the solution is stable up to 1 month at 4°C unless turbidity develops; can also be atored at −20°C for 6 months.

4. Enzymes – Dye reagent mix:

To 100 ml of the enzyme solution, add 1 ml of dye solution and mix. Prepare fresh everyday but the solution is stable up to one month at 4°C.

PROCEDURE

Pipette out into a series of test tubes 0 – 0.5 ml of standard glucose solution and make the final volume to 0.5 ml. To each tube add 5.0 ml of Enzymes and dye reagent, mix and incubate at 37° C for 30 minutes (90% of the reaction is completed in 15 minutes at 37°C). (Avoid exposure to direct sunlight or bright daylight). Add 0.1 ml of 4 N HCl to each tube, mix, incubate for at least 5 minutes and measure the absorbance at 420 nm. (The absorbency can be measured at 425 nm directly without adding HCl. But the readings should be completed within 30 minutes).

INTERFERING SUBSTANCES

Ascorbic acid, catechols, cysteine, glutathione, acetyl-salicylic acid, L-Dihydroxy phenylalanine (L-DOPA), mercurial diuretics and tetracycline.

Note:

Glasswares can be cleaned free of stain by rinsing with acetone and then washing with a detergent.

A NOTE ON THE ESTIMATION OF BLOOD GLUCOSE

Glucose can be determined in whole blood, serum and plasma. Whole blood should be taken directly in a tube containing heparin (an anticoagulant) and sodium fluoride. (to avoid loss of glucose by glycolysis). At room temperature the blood glucose undergoes glycolysis at a rate of 5% per hour; fluoridated blood is stable for several hours at room temperature. For preparation of plasma, the whole blood is centrifuged for 10 minute at 5000 rpm to sediment the red cells. The supernatant is the plasma.

For preparation of serum, blood is drawn into a tube without any anticoagulant and fluoride and allowed to clot for about 30 min at RT for 30 min at 4°C. The serum is obtained from the clot by low speed centrifugation (1000 rpm x 10 min). (Please note, about 5% of the glucose will be lost by glycolysis for each hour that the serum remains with the clot). Once separated from the clot, the glucose level of serum remains constant for at least 24 hours even at room temperature. (25 ìl of either plasma or serum can be used as such for glucose estimation).

NORMAL VALUES FOR FASTING ADULTS

Whole blood	-	50–90 mg/100ml
Serum or plasma	-	65–100 mg/100ml

(Blood sugar is increased in diabetes mellitus, decreased in hyper-insulinism, Addison's disease and glycogen storage disease).

44

ESTIMATION OF NUCLEIC ACID

Nucleic acids are present in all living cells and are responsible for storage and transmission of genetic information. Chemically nucleic acids are polymers of nucleotides. The nucleotides contain a nitrogenous base (either a purine or a pyrimidine), a pentose sugar (either deoxyribose in the case of DNA's or ribose in the case of RNA's) and a phosphate group on the sugar. The nucleotides are connected by the phosphate groups, which are known as phosphodiester bonds. Nucleic acids are quickly and conveniently estimated by spectrophotometric methods. Colorimetric methods are also used to estimate DNAs and RNAs in crude preparation.

ESTIMATION OF DNA BY DIPHENYLAMINE METHOD

This method can be used relatively crude extracts where direct measurements of DNA by ultraviolet absorbency are not possible. The assay in specific for deoxyribose, although very high concentrations of ribose or sucrose must be avoided.

PRINCIPLE

In the presence of strong acid, the deoxyribose moieties of DNA from hydroxylevulinicacid. The hydroxylevulinic acid reacts with diphenylamine and produces a blue coloured complex as shown in Figure (next page) with a $\ddot{e}_{max}$ of 595 nm. The colour is measured at 585 nm or using a red filter. It is important to note that only the deoxy moieties of purine nucleotides (adenine and guanine) react and produce the colour.

REAGENTS

1. Preparation of Diphenylamine reagent

Dissolve 1.0 g diphenylamine (AR grade) in 97.5 ml of glacial acetic acid and then add 2.5 ml of conc. Sulphuric acid. Mix well. Store the reagent in a brown bottle at room temperature (this reagent is stable for months)

2. DNA stock solution

Dissolve calf thymus DNA or equivalent at 0.2mg/ml in 5mM NaOH. Store at 4°C (This reagent is stable for at least 6 months)

PROCEDURE

1. Pipette 0.05 ml to 1.0 ml DNA stock solution into clean tubes and make the volume to 1.0 ml with distilled water as shown in Table 9.1 below. Mix them well.
2. Add 5 ml of diphenylamine reagent to each tube and vortex. Cover the tubes with aluminum foil and secure it with rubber bands.
3. Now place the tubes in a boiling water bath for 10 minutes.
4. Cool the tubes to room temperature and read the absorbency at 595 nm (or using a red filter).
5. Draw a standard graph using the values obtained for A_{595} as function of DNA concentration.

Note: The diphenylamine reagent is not water-soluble. Therefore, rinse the cuvettes, glasswares and test tubes in ethanol before washing them in water.

	0	1	2	3	4	5	6	7	8	9
DNA (0.2mg/ml)	0	0.05	0.1	0.2	0.3	0.4	0.5	0.6	0.8	1
dH2O	1	0.95	0.9	0.8	0.7	0.6	0.5	0.4	0.3	0
mg?of DNA	0	10	20	40	60	80	100	120	150	200
DPA Rgt 5ml										

DPA Rgt ◄————————————————— 5 ml —————————————————►

Vortex and place the tubes in boiling water bath for 10 min and read the absorbency at 595 nm.

THE ISOLATION OF DNA FROM SPLEEN

PRINCIPLE

All cells contain DNA but some tissues may contain small amount of DNA while others may have high deoxyribcnuclease activity that may cause fragmentation of the DNA. Hence it is advisable to choose such a sample that should contain DNA in high quantity and low deoxyribonuclease activity. In this respect lymphoid tissue is a very good source, like thymus and spleen.

The tissue is homogenized with isotonic saline solution (pH 8.4) containing sodium citrate as buffer. The deoxyribonucleoprotein is insoluble at this ionic strength and separates well from other proteins. Deoxyribonuclease activity is inhibited by Sodium citrate, which binds to Ca^{++} and Mg^{++} (cofactors for this enzyme). Cold extraction is preferred to DNA'ase activity. Degradation of the DNA is avoided by the use of glass or plastic vessels.

Addition of ethanol will finally cause precipitation of DNA as a fibrous white mass. DNA is dissolved in saline buffered (pH 7.4) with sodium citrate. It can be stored in deep freezer for several months. Avoid drying.

MATERIALS

- Spleen.
- Buffered saline (0.15 mol/L Nail buffered with 0.015 mol/L Sodium citrate, pH 7).
- Sodium Chloride 2M.
- Ethanol & Ether.
- Solvents used:

    1. 70 per cent v/v ethanol,
    2. 80 per cent v/v ethanol,

3. absolute ethanol and
4. Ether.

METHOD

- Take small pieces of spleen, homogenize with 20 ml of buffered saline for 1 min.
- Centrifuge at 5000 g for 15 min and rehomogenize the precipitate in a further 40 ml of buffered saline.
- Discard the supernatant and dissolve the combined sediments uniformly in 2 mol/litre NaCl to a final volume of 100ml
- Add an equal volume of ice-cold water stirring the solution continuously with a glass rod.
- Spool the fibrous precipitate on to a glass rod and leave it to stand in a beaker for 30 min.
- During this time the clot will shrink and the liquid expressed should be removed with filter paper.
- Dissolve the deoxyribonucleoprotein in about 100ml of 2 mol/litre NaCl, add an equal volume of the chloroform/amyl alcohol mixture (6:1), and blend for 30s.
- Centrifuge the emulsion at 5000 g for 10-15 min and collect the upper (opalescent) aqueous layer containing the DNA with the help of a micropipette to avoid disturbance of the denatured protein at the interface of the two liquids.
- Repeat the process twice and collect the supernatant in a 500 ml beaker.
- Precipitate the DNA by slowly stirring 2 volumes of ice-cold ethanol with the supernatant and collect the mass of fibres on the glass stirring rod.
- Carefully remove the rod and gently press the fibrous DNA against the side of the beaker to expel the solvent.
- Remove the last traces of ether by standing the DNA in a fume cupboard for about 10 min.
- Weigh the dry DNA and dissolve by continuously stirring in buffered saline diluted one in ten with distilled water (2 mg/ml); store frozen until required.

45

1. ORCINOL METHOD

The method can be used on relatively crude preparations of RNA where direct measurements of RNA by ultraviolet absorbency are not possible.

PRINCIPLE

Ribose moieties present in ribonucleic acids react with hot hydrochloric acid and produce furfural. The furfural thus formed is reacted with orcinol in presence of ferric ions to give a brilliant green colour as shown below in Figure. The intensity of the colour is measured at 665 nm. (The purine nucleotides are generally more reactive than pyrimidine nucleotides).

REAGENTS

1. **(a) 1% orcinol solution**
   Dissolve 1 g orcinol in 5 ml ethanol and make the volume to 100 ml with distilled water. It can be stored in the refrigerator for several months.
   (b) Concentrated hydrochloric acid
   (c) 10% Ferric chloride solution (prepare afresh)
   Dissolve 2 g of $FeCl_3.6H_2O$ in distilled water and make up the volume at 20 ml.
2. **Preparation of Orcinol reagent: (Prepare afresh)**
   Mix 10 ml of 10% Ferric chloride solution to 390 ml of conc. HCl. Add the mixture slowly into the 100 ml 1% orcinol solution with stirring. The final volume is 500 ml.

RNA stock solution

Dissolve yeast RNA or equivalent at 0.1 mg/ml in distilled water (store at -20°C).

OH   CH₂OH   CHO

Conc. HCL →

OH OH

Ribofuranose   Furfural

CH3

CHO   FeCl3 (max 665 nm) → Green Colour Complex

HO   OH

Furfural   Oricinol

PROCEDURE

1. Pipette out 0.05 ml to 2.0 ml of RNA stock solution into the clean test tubes and make up the final volume to 2.0 ml with distilled water as shown in the Table 9.2 below and mix.
2. Add 4.0 ml of the orcinol reagent to each tube and vortex.
3. Cover the tubes with aluminum foil, secure with a rubber band and place the tubes in a boiling water bath for 15 minutes.
4. Cool the tubes in a running tap water to room temperature and measure the absorbency at 665 nm (or using red filter).
5. Draw a standard graph by plotting the values as a function of RNA concentration and express the results as ribose equivalent.

Table Orcinol Reaction protocol

	0	1	2	3	4	5	6	7	8	9	10
RNA (0.2mg/ml)	0	0.05	0.1	0.2	0.3	0.4	0.5	0.75	1	1.5	2
dH2O	2	1.95	1.9	1.8	1.7	1.6	1.5	1.25	1	0.5	0
mg RNA	0	5	10	20	30	40	50	75	100	150	200

Vortex and place the tubes in boiling water bath for 15 min and read the absorbency at 655 nm.

Estimation of Nucleic Acids by UV method

This is a quick, simple and routine procedure for estimation of relatively pure nucleic acid. This is based on the fact, that conjugated double bonds in nitrogenous base of nucleic acids absorb light strongly at 260 nm. Based on the amount of UV light absorbed at 260 nm, the nucleic acid concentrations are calculated as follows:

$1\ A_{260}$ unit of duplex nucleic acid (DNA) $= 50.0$ mg/ml
$1\ A_{260}$ unit of single standard nucleic acid DNA $= 36.5$ mg/ml
$1\ A_{260}$ unit of single standard nucleic acid RNA $= 40.0$ mg/ml
$1\ A_{200}$ unit of oligonucleotides $= 20.0$ mg/ml

The 260/280 – absorbency ratio also indicates the purity of DNA or RNA preparations, e.g.,

The 260/280 ratio for pure DNA $= 1.8$
The 260/280 ratio for pure RNA $= 2.0$

Unlike calorimetric methods, the UV method requires relatively pure preparations of nucleic acids for estimations.

PROCEDURE

Step 1

The crude extract of cells can be prepared by various methods such as grinding of frozen cells of bacteria, yeast and fungi or by protoplast preparation of such above cells, using cell wall degrading enzymes and then lysing the cells. (The bacterial cell extract can also be prepared by sonication. The yeast cells can also be broken open by vortexing vigorously with acid washed glass beads (~500 ì size) for 1-2 min). The plant cell extracts are prepared routinely by grinding the plant materials in liquid nitrogen. The animal cell extracts are prepared by Potter Elvehjm homogenizers or by blending in a mixie. (For preparation of cell extracts appropriate buffer system should be used, especially to inactivate nucleases).

Step 2

In the second step, the crude homogenate is centrifuged at 12,000 rpm fro ~ 20 min at 4°C.

Step 3

In the third step, the clear homogenate is extracted with equal volume of buffer saturated phenol*, then with a mixture of phenol: chloroform: isoamyl alcohol** (25:24:1 v/v) and finally with chloroform: isoamyl (24:1 v/v).

Step 4

The final aqueous (top) layer is carefully pipetted out into a clean tube without disturbing the interphase (made up of mainly denatured proteins). The nucleic acid is precipitated with two volumes of cold alcohol and 0.1M NaCl at 0° C or –20 ° C for ~ 30 min. The precipitate is collected by centrifugation at 12,000 rpm for 30 min at 4° C, air-dried and dissolved in TE (10 mM EDTA). Now this preparation can be used to estimate concentration of nucleic acid by UV method.

*Take required volume phenol. Add equal of 0.1M Tris, pH.8.0 Mix by vortexing and allow it to stand for 5 min. Two phases separate. Pipette out and discard the top aqueous layer and repeat the extraction. The bottom phenol phase obtained thus after two extractions with Tris is used. (Phenol is acidic and hence is saturated with Tris buffer).

**Phenol denatures and precipitates proteins very efficiently. Chloroform also denatures proteins but also helps in quicker phase separation. In the second extractions it removes excess phenol which otherwise might interfere with subsequent enzymes digestion steps. Isoamyl alcohol prevents frothing and also helps in phase separation.

Note

RNAs from the above preparation can be removed by treating the sample with DNase free RNase and then extracting again with phenol, phenol: chloroform and chloroform as described above. The DNA in the final aqueous phase is precipitated as described above and estimated.

46

EXTRACTION OF LIPID

AIM OF THE EXPERIMENT

To extract and quantify lipids from oil seeds

MATERIALS REQUIRED

A. Chemicals

- Chloroform
- Methanol (2:1)

B. Glasswares

- Small beaker
- Measuring cylinder (50ml & 100 ml)
- Micropipette
- Funnel and stand

C. Micellaneous Articles

- Mortal and pestle
- Filter paper
- Water bath

D. Balance

E. Specimen: Oil seed or any other oil bearing specimen

PROCEDURE

1. Weigh 1g of seeds after removing the surface coat (or any other oil bearing specimen) and homogenize with chloroform: methanol reagent at the ratio of 2:1. Adjust the homogenate volume to 20ml.
2. Filter the homogenate and collect the filtrate in a beaker.
3. To the filtrate add water, 20% of its volume (i.e, 4ml for 20ml extract) and let the oil be separated into two phases by allowing it to stand for 10 min.

4. Remove the upper phase, which is mostly composed of water and reagent, with help of a micropipette
5. Add methanol to the lower phase so that it mixes into one phase
6. Evaporate the lower phase to dryness in a water bath. The lower phase now contain only oil
7. Calculate the gram % of oil of the seed taken

PRECAUTION

The solvent taken in this experiment are highly inflammable hence maximum care should be taken while handling them. Keep them away from from open flame.

47

ASSAY OF AMYLASE ACTIVITY

AIM OF THE EXPERIMENT

To demonstrate the amylase activity in human saliva

MATERIALS REQUIRED

A. Chemicals

- 005M Iodine in 3% KI (potassium iodide)
- 1% NaCl solution
- Starch

B. Glasswares and Miscellaneous Articles

- Small beaker
- Test Tubes
- Micropipette
- Rubber band

C. Specimen

Human saliva can be obtained by chewing a rubber band. The saliva is collected in a beaker

PROCEDURE

1. Dilute the saliva to 1:20 with distilled water
2. Prepare 0.005N iodine solution in 3% KI
3. Weigh 1g of starch and mix well with 100ml of distilled water, boil and cool
4. Pipette out 5ml of starch solution in two test tubes
5. Add 1ml of 1% NaCl solution, mix and keep at 37° C
6. Add 1ml of dilute saliva in one of the test tube, while the other test tube will be the control
7. To the test tubes add a few drops of the iodine solution

OBSERVATION

The control tube, which do not cantain saliva, will be the refernce for the starch in the tubes and reaction at different intervals. While the control will have blue colour the experimental solution will be colourless.

SEPARATION AND ANALYTICAL TECHNIQUES

48

The principle of Electrophoresis involves the motion of a particle in a fluid medium, where the gravitational attraction (on a flowing liquid) is the sedimenting force, and the frictional force is the opposing force and is proportional to the velocity of the particle. These forces balance each other acting in opposite directions and hence balance each other. This will lead to the uniform motion of the particle in the mobile liquid medium (i.e. the particle moves at constant velocity). In case the external field is an electric field instead of a gravitational field, a macromolecule will respond to the external field in two ways. In case the molecule is charged, it will migrate in an electric field to the electrode of opposite charge. Under this principle the technique of electrophoresis functions.

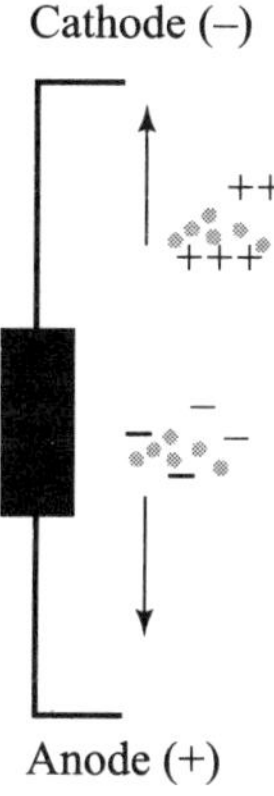

Fig. 41   Separation of charged particles by electric current

Consider the simple case of a charged particle ($+Q$) moving in an electric field (E) in a non conducting medium, such as water. The net force F_{tot} on the particle will be zero if the particle is moving at a constant velocity towards the cathode (since $F=ma$, and the acceleration (a) of the particle is 0 at constant velocity). Two forces act on the particle,

1. F_E, the force exerted on the charged particle by the field, in the direction of the motion (i.e., toward the cathode), and
2. F_f, the frictional force on the charged particle in the direction opposite to the motion (toward the anode ($+$) electrode), which tries to inhibit its motion toward the cathode. This is shown in the diagram below:

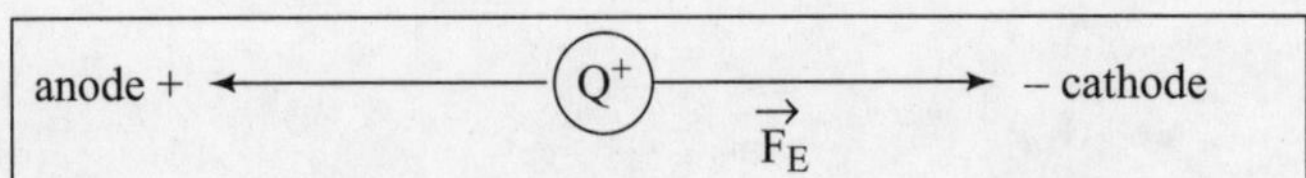

Therefore:
(1) F_{tot} = $F_e + F_f = 0$, where
(2) F_e = QE (the electric force) and
(3) F_f = -fv (the frictional force),

where:

- v is the velocity of the particle, and
- f is a constant called the frictional coefficient.

Equation (3) reveals that the frictional force (F_f) is proportional to the velocity of the particle and the frictional coefficient is dependent upon the size and shape of the molecule. Larger the molecule larger will be the frictional coefficient (i.e. more resistance to motion of the molecule). Thus, all electrophoresis operations are governed by the principle:

$$\text{Molecule a of mobility} = \frac{(\text{Voltage applied})(\text{Molecule on the charge net})}{(\text{molecular friction})}$$

Modern day electrophoresis is conducted in solid gels (such as agarose or polyacrylamide). The solid gel is porous to solute and solvent molecules and serves as a medium for electrophoresis while helping to eliminate convection forces in the liquid which would interfere with the separation.

It has been determined that the actual electrophoretic mobility of the protein, U, is a function of the mobility of the protein in a concentrated sucrose solution (Uo) and T, the total concentration of the agarose / acrylamide in the polymerized gel. The higher the concentration of agarose/acrylamide in the unpolymerized gel solution, the smaller the size of the pores in the polymerized gel.

The one having the more elongated shape would have a lower electrophoretic mobility. Hence both electrophoretic mobility and sieving effects would cause this protein to run anomalously slow and have a higher apparent molecular weight. Also imagine two globular proteins of different size but with compensatory charge differences which might allow the two proteins to migrate at the same speed in the gel.

Fig. 42   An Electrophoresis unit with power supply showing gel cast with separated protein bands

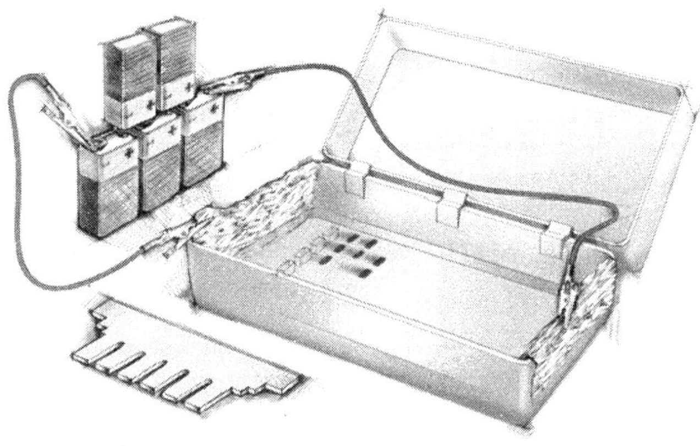

Fig. 43   A cheap home made Electrophoresis unit

ADVANTAGES OF A DISCONTINUOUS GEL SYSTEM

The main advantage is that the proteins electrophorese quickly through the stacking gel and "stack" at the interface between the two gels, before they enter the gels. This increases the compactness of the proteins before they enter the running gel and increases resolution.

DETECTION OF PROTEINS IN THE GEL

Most proteins are unable to absorb visible wavelengths of light, and hence are not visible during electrophoresis. To ascertain that the proteins are not eluted from the gel into the lower buffer reservoir, an anionic dye, called bromophenol blue, is added to the protein before it is placed on the gel. The electrophoresis is halted when the dye reaches near the end of the running gel.

STAINING TECHNIQUES

To detect proteins in the gel following techniques can be used:

1. Staining with Coomassie Blue

The most common is to stain the gel with Coomassie Blue, dissolved in a methanol/acetic acid solution. Proteins bind Coomassie Blue. To fix the dye to the gel, methanol and acetic acid is added to prevent its diffusion into the solution. The background stain in the gel is removed with acetic acid/methanol, leaving the blue colored protein bands. Some proteins, however, will not be stained with Coomassie blue.

2. Silver Staining

It is another common staining technique, which involves the reduction of Ag(I) to elemental silver and its deposition by protein in the appropriate reaction solutions, much as in a photographic process. A developer and fixer solution is required. The technique is much more sensitive than Coomassie Blue staining.

VARIATIONS ON GEL ELECTROPHORESIS

Isoelectric focusing

In this technique, a pH gradient is set up within the polyacrylamide gel. This is accomplished by pre-electrophoresing a series of low molecular weight molecules containing amino and carboxyl groups called ampholytes. When subjected to an electric field, the most negative of the species will concentrate at the anode, while the most positive will concentrate toward the cathode, as shown in the figure below. The remaining ampholytes will migrate in-between, with the net effect being that the ampholytes migrate to their isoelectric point and set up a linear pH gradient in the gel. A protein applied to the gel will migrate to the pH corresponding to its isoelectric point and stop.

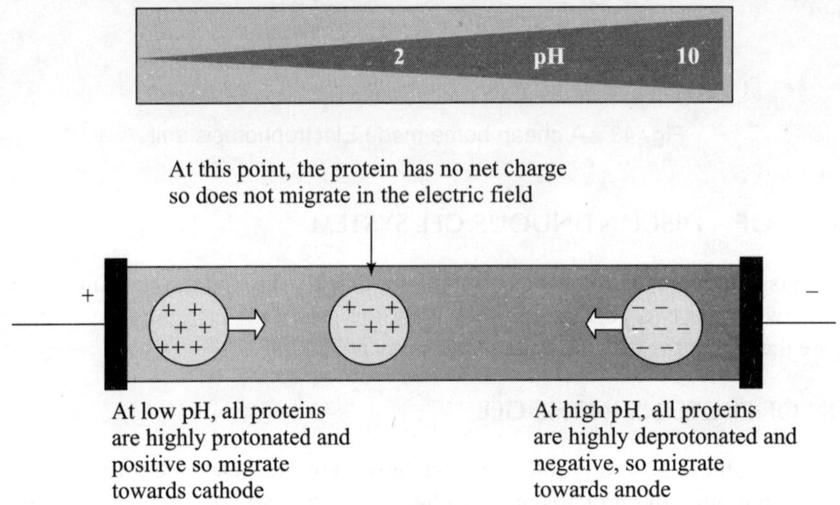

Fig. 44   Isoelectric Focussing

2D ELECTROPHORESIS

This typically involves subjecting the proteins to isoelectric focusing electrophoresis in a polyacrylamide gel cast in a narrow cylindrical tube. After this electrophoresis, the tube gel is removed, and placed across the top of the stacking gel and subjected to SDS-polyacrylamide gel electrophoresis in a direction 90o from the initial isoelectric focusing experiment. If the proteins were derived from cells labeled with [35]Met,   representing unique proteins can be obtained from a given cell population.

WESTERN BLOTTING

After a standard SDS-slab electrophoresis experiment is run, the gel is overlaid with a piece of nitrocellulose filter paper. The sandwich of gel and filter paper is placed back into an electrophoresis chamber, such that the proteins migrate from the gel into the nitrocellulose, where they irreversibly bind. The filter paper can be removed and soaked in a solution containing a specific antibody to a protein of interest on the nitrocellulose. This protein- antibody complex on the filter paper can then be detected by adding a fluorescently-labeled antibody that binds the first antibody, for instance.

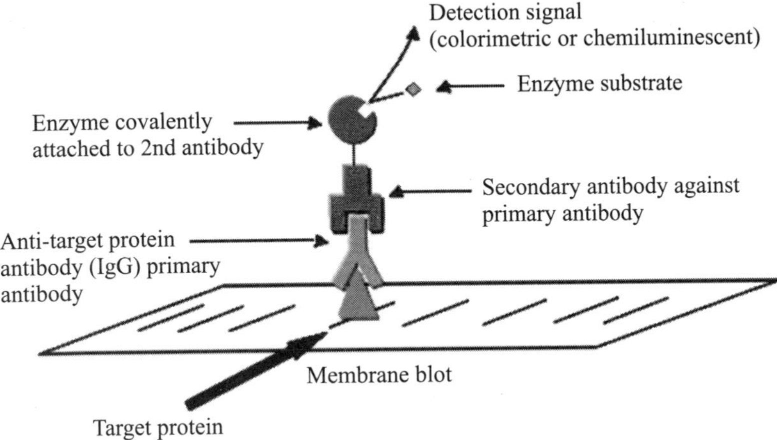

Fig. 45 Detection in Western Blots

49

AGAROSE GEL ELECTROPHORESIS

Agarose gel electrophoresis is the easiest and commonest way of separating and analyzing DNA. The purpose of the gel might be to look at the DNA, to quantify it or to isolate a particular band. The DNA is visualised in the gel by addition of ethidium bromide. This binds strongly to DNA by intercalating between the bases and is fluorescent meaning that it absorbs invisible UV light and transmits the energy as visible orange light.

The standard method used to separate, identify, and purify DNA fragments are electrophoresis through agarose gels. The technique is simple, rapid to perform, and capable of resolving mixtures of DNA fragment that cannot be separated adequately by other sizing procedures. Furthermore, the location of DNA within the gel can be determined directly: Bands of DNA in the gel are stained with the intercalating dye ethidium bromide; as little as 1 ng of DNA can be detected by direct examination of the gel in ultraviolet light.

Agarose, which is extracted from seaweed, is a linear polymer whose basic structure is shown in figure.

Fig. 46 A red alga, the source of agarose

Agarose structure unit

MATERIALS

- Electrophoresis Chamber
- Agarose
- Water
- Erlenmeyer flask
- Graduate cylinder
- Aluminum foil or plastic wrap
- Balance
- Hot plate or Microwave
- Paper towels
- Masking Tape
- Electrophoresis Buffer
- Weigh boat or wax paper
- Latex Gloves
- Goggles
- Procedure

MAKING THE GEL

To measure the volume, multiply the width, height, and length of the gel form. When you measure the height, take into consideration where the agarose should come up to on the comb used to create the wells.)

- If the volume of the gel form is 100 ml then 1 gram of agarose is needed to make a 1% gel.
- Take the amount of agarose on wax paper and pour it into a flask.
- Prepare the gel form by taping the ends with masking tape. Be careful that the gel form is sealed well; otherwise, the agarose solution will leak onto your work surface before it has a chance to solidify.
- Measure the volume of TBE buffer to dissolve the agarose into a 1% solution. (**Note**: For making 50X electrophoresis buffer, dilute the solution in the following manner: 6.25 ml of 50X buffer with 307.75 ml of distilled water to yield a total volume of 312 ml of dilute buffer. Use this buffer to fill the electrophoresis chamber).
- Place the contents of the flask on a hotplate.
- Gently and constantly swirl the contents until the agarose is completely dissolved. You can check to see if the agarose is in solution by holding the flask up to the light and looking for floating particles. Your solution should be clear.
- Allow the solution to cool to 50-55°C or until you can place your hand on the flask. If your solution begins to solidify before you have a chance to pour it, reheat the solution on the hotplate.
- Place the comb in the gel form.
- Pour the agarose solution into the gel form to where is comes a little over halfway up the comb.
- Allow the solution to set (solidify) undisturbed. It should look like milky jello.

PREPARING THE SAMPLES

Make the loading buffer as instructed on attached sheet. Why is glycerol used in the loading buffer? Mix your samples with the loading buffer.

LOADING THE GEL

- Remove the comb and place the gel form into the electrophoresis chamber filled with 1X TBE buffer. The buffer should cover the gel.
- Note the lane that each sample is placed in.
- Use a clean micropipet tip or Pasteur pipet to load each sample.
- Using both hands to steady the pipet tip, position the tip over the well and gently expel the sample into the well. Too much pressure will flush the sample out of the well and into the buffer. Be careful not to puncture the bottom of the gel; otherwise, your sample will leak out of the bottom.

Fig. 47   A horizontal Gel System used in agarose electrophoresis showing gel casting, loading of sample and power supply

RUNNING THE GEL

- Once all the samples have been loaded into the wells, place the cover over the electrophoresis chamber.
- Attach the electrodes to the electrophoresis chamber and then to the power supply.
- Place the cover over the electrophoresis apparatus.
- Never put your fingers into the naked terminals on the power box or into the buffer. There will be danger of getting electrocuted.
- Run the gel at 50-100 volts for approximately 60-95 minutes.
- When the DNA fragments have been separated, turn off the power supply.
- Unplug all the electrodes.
- With gloved hands, remove the gel form from the electrophoresis chamber to stain the gel.

STAINING AND VISUALIZING THE GEL

- Make a dilute solution of Methylene Blue plus Staining Solution by mixing 60 ml of concentrated staining solution with 540 ml of distilled water.
- Take the gel and place into a rectangular plastic container with gloved hands. Remove the gel from the electrophoresis chamber. Staining the gel in the chamber will ruin the electrophoresis apparatus.

- Pour the diluted staining solution over the gel until the gel is completely submersed.
- Stain the gel for 20-30 minutes. Your gel should appear as a dark blue piece of jello.
- Prior to de-staining, pour the staining solution into a glass bottle and store at room temperature. You can reuse this solution for up to 5 times.
- To de-stain the gel, submerse the gel in distilled water. You may need to pour off the water and add fresh water to the container as the stain leaches out from the gel. If the gel is too dark, de-stain overnight. If the gel is too light, repeat the staining process.

Note: Agarose gels can be sealed in plastic wrap and stored in the refrigerator for several weeks. When gels are no longer needed, dispose of them in the trash.

50

DNA GEL ELECTROPHORESIS

Agarose gel electrophoresis can be used to separate DNA molecules according to size. The procedure is the same as when we separated the food dyes. In fact, we load a dye with our DNA samples so that we can monitor the progress of the electrophoresis. The dye solution also contains glycerol to provide a dense mixture that will stay in the bottom of the well.

PREPARING AND RUNNING STANDARD AGAROSE DNA GELS

The equipment and supplies necessary for conducting agarose gel electrophoresis are relatively simple and include:

- An electrophoresis chamber and power supply
- Gel casting trays, which are available in a variety of sizes and composed of UV-transparent plastic. The open ends of the trays are closed with tape while the gel is being cast, then removed prior to electrophoresis.
- Sample combs, around which molten agarose is poured to form sample wells in the gel.
- Electrophoresis buffer, usually Tris-acetate-EDTA (TAE) or Tris-borate-EDTA (TBE).

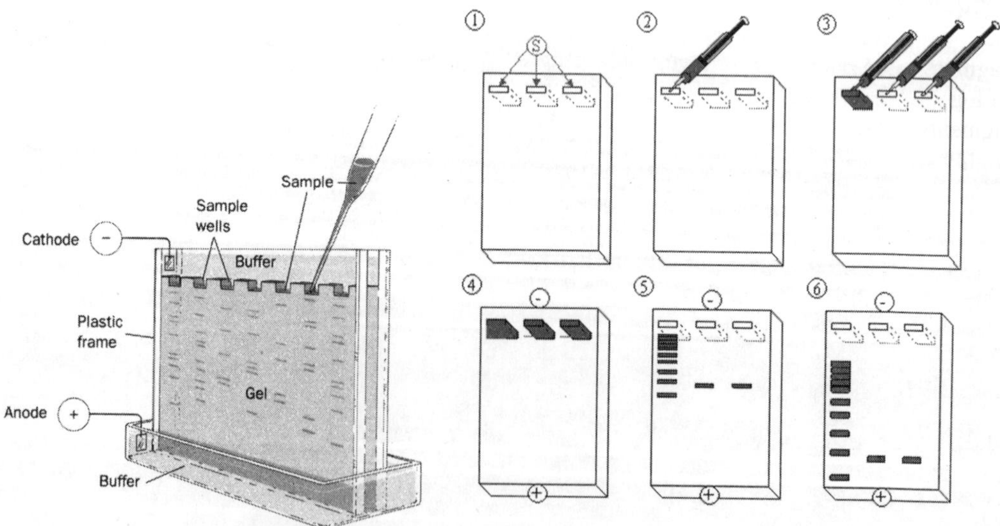

Fig. 48 Process of DNA-Gel electrophoresis

- Loading buffer, which contains something dense (e.g. glycerol) to allow the sample to "fall" into the sample wells, and one or two tracking dyes, which migrate in the gel and allow visual monitoring or how far the electrophoresis has proceeded.
- Ethidium bromide, a fluorescent dye used for staining nucleic acids. *NOTE: Ethidium bromide is a known mutagen and should be handled as a hazardous chemical - wear gloves while handling.*
- Transilluminator (an ultraviolet lightbox), which is used to visualize ethidium bromide-stained DNA in gels. *NOTE: always wear protective eyewear when observing DNA on a transilluminator to prevent damage to the eyes from UV light.*

To pour a gel, agarose powder is mixed with electrophoresis buffer to the desired concentration, then heated in a microwave oven until completely melted. Most commonly, ethidium bromide is added to the gel (final concentration 0.5 ug/ml) at this point to facilitate visualization of DNA after electrophoresis. After cooling the solution to about 60C, it is poured into a casting tray containing a sample comb and allowed to solidify at room temperature or, if you are in a big hurry, in a refrigerator.

After the gel has solidified, the comb is removed, using care not to rip the bottom of the wells. The gel, still in its plastic tray, is inserted horizontally into the electrophoresis chamber and just covered with buffer. Samples containing DNA mixed with loading buffer are then pipeted into the sample wells, the lid and power leads are placed on the apparatus, and a current is applied. You can confirm that current is flowing by observing bubbles coming off the electrodes. DNA will migrate towards the positive electrode, which is usually colored red.

The distance DNA has migrated in the gel can be judged by visually monitoring migration of the tracking dyes. Bromophenol blue and xylene cyanol dyes migrate through agarose gels at roughly the same rate as double-stranded DNA fragments of 300 and 4000 bp, respectively.

When adequate migration has occured, DNA fragments are visualized by staining with ethidium bromide. This fluorescent dye intercalates between bases of DNA and RNA. It is often incorporated into the gel so that staining occurs during electrophoresis, but the gel can also be stained after electrophoresis by soaking in a dilute solution of ethidium bromide. To visualize DNA or RNA, the gel is placed on a ultraviolet transilluminator. Be aware that DNA will diffuse within the gel over time, and examination or photography should take place shortly after cessation of electrophoresis.

MIGRATION OF DNA FRAGMENTS IN AGAROSE

Fragments of linear DNA migrate through agarose gels with a mobility that is inversely proportional to the $\log_{10}$ of their molecular weight. In other words, if you plot the distance from the well that DNA fragments have migrated against the $\log_{10}$ of either their molecular weights or number of base pairs, a roughly straight line will appear.

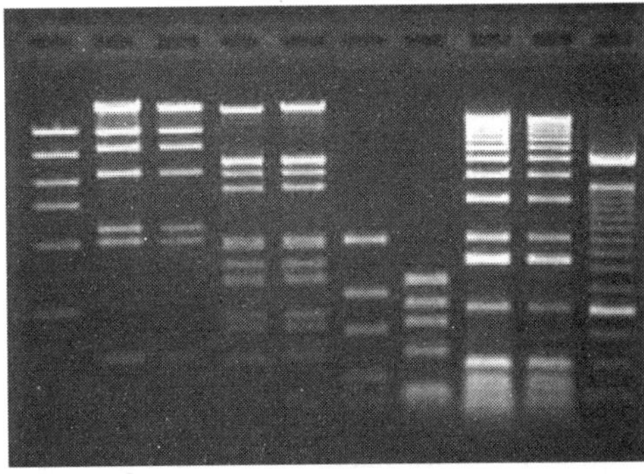

Fig. 49 A DNA agarose electrophoresis gel observed under UV light, The DNA has been stained with ethidium bromide, which makes DNA fluoresce under UV.

Circular forms of DNA migrate in agarose distinctly differently from linear DNAs of the same mass. Typically, uncut plasmids will appear to migrate more rapidly than the same plasmid when linearized. Additionally, most preparations of uncut plasmid contain at least two topologically-different forms of DNA, corresponding to supercoiled forms and nicked circles. The image to the right shows an ethidium-stained gel with uncut plasmid in the left lane and the same plasmid linearized at a single site in the right lane.

Several additional factors have important effects on the mobility of DNA fragments in agarose gels, and can be used to your advantage in optimizing separation of DNA fragments. Chief among these factors are:

1. Agarose Concentration

By using gels with different concentrations of agarose, one can resolve different sizes of DNA fragments. Higher concentrations of agarose facilite separation of small DNAs, while low agarose concentrations allow resolution of larger DNAs.

Range and Agarose concentration

Agarose concentration (%)	Separation range (Kb)
0.3	5 to 60
0.6	1 to 20
0.8	0.8 to 10
1.0	0.4 to 8
1.2	0.3 to 6
1.5	0.2 to 4
2.0	0.1 to 3

The figure below shows migration of a set of DNA fragments in three concentrations of agarose, all of which were in the same gel tray and electrophoresed at the same voltage and for identical times. Notice how the larger fragments are much better resolved in the 0.7% gel, while the small fragments separated best in 1.5% agarose. The 1000 bp fragment is indicated in each lane.

2. Voltage

As the voltage applied to a gel is increased, larger fragments migrate proportionally faster than small fragments. For that reason, the best resolution of fragments larger than about 2 kb is attained by applying no more than 5 volts per cm to the gel (the cm value is the distance between the two electrodes, not the length of the gel).

3. Electrophoresis Buffer

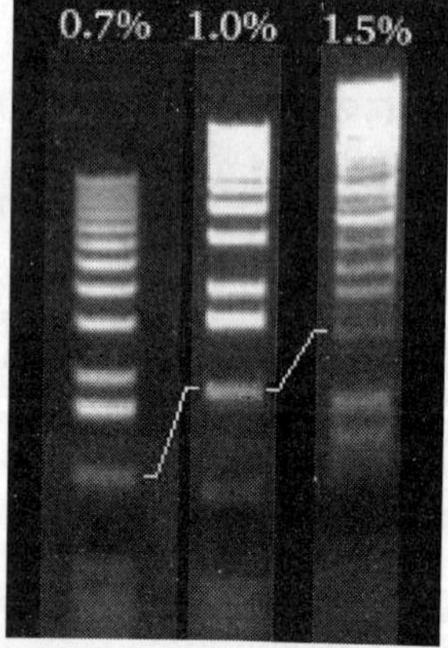

Fig. 50 DNA agarose Gel in three concentration of agarose

Several different buffers have been recommended for electrophoresis of DNA. The most commonly used for duplex DNA are TAE (Tris-acetate-EDTA) and TBE (Tris-borate-EDTA). DNA fragments will migrate at somewhat different rates in these two buffers due to differences in ionic strength. Buffers not only establish a pH, but provide ions to support conductivity. If you mistakenly use water instead of buffer, there will be essentially no migration of DNA in the gel! Conversely, if you use concentrated buffer (e.g. a 10X stock solution), enough heat may be generated in the gel to melt it.

4. Effects of Ethidium Bromide

Ethidium bromide is a fluorescent dye that intercalates between bases of nucleic acids and allows very convenient detection of DNA fragments in gels, as shown by all the images on this page. As described above, it can be incorporated into agarose gels, or added to samples of DNA before loading to enable visualization of the fragments within the gel. As might be expected, binding of ethidium bromide to DNA alters its mass and rigidity, and therefore its mobility.

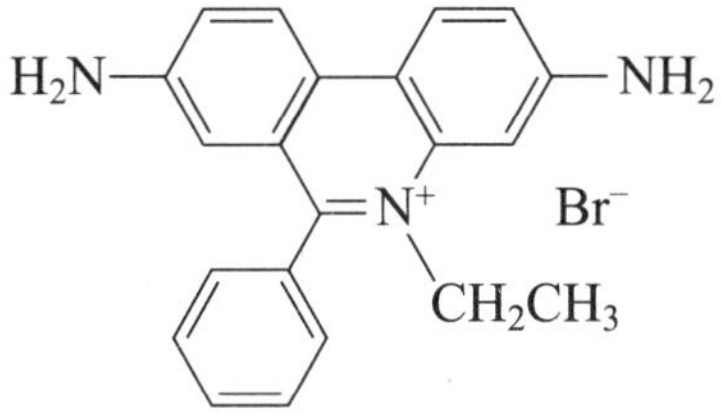

Structure of Ethidium Bromide

OTHER CONSIDERATIONS

- Agarose gels, as discussed above provide the most commonly-used means of isolating and purifying fragments of DNA, which is a prerequisite for building any type of recombinant DNA molecule.
- By varying buffer composition and running conditions, the utility of agarose gels can be extended. Examples include:
- Pulsed field electrophoresis is a technique in which the direction of current flow in the electrophoresis chamber is periodically altered. This allows fractionation of pieces of DNA ranging from 50,000 to 5 millon bp, which is much larger than can be resolved on standard gels.
- Alkaline agarose gels are prepared with and electrophoresed in buffers containing sodium hydroxide. Such alkaline conditions are useful for analyzing single-stranded DNA

MATERIAL

DNA samples
DNA size standard
Loading dye solution
Agarose
TBE buffer
Sybr Green I solution diluted 10,000-fold
Dark Reader light box
Plastic wrap

PROCEDURE

(1) Dissolve 1 g of agarose in 100 ml of 1X TAE or TBE buffer (gives a 1% gel).
(2) Cast the gel with the comb in place.
(3) Add 6X gel loading buffer to sample and load the samples into wells.
(4) Run the gel submerged in 1X TAE or TBE (30-60 min. at 100-150V)
(5) Stain the gel with ethidium bromide solution (10 µl per 100ml of buffer) for 10-15 min.
(6) View on a UV transilluminator.

INFERENCE

(a) TAE is better for cloning work as the borate in TBE may affect subsequent ligation reaction if it is not removed well.

(b) The percentage of gel used depends on the size of DNA to be separated. Low percentage gel separates high m.w. DNA while high percentage gel separates low m.w DNA.

(c) Gel may be cast with ethidium bromide solution added, this will save time staining but the gel will also run slightly slower (note that ethidium bromide is a powerful mutagen, so be careful in its use).

(d) In general the supercoiled circular plasmid runs faster than cut plasmid. Multiple bands can be seen with closed circular plasmid (supercoiled DNA, relaxed DNA, denatured DNA, catenae, etc.).

(e) The distance tranvelled in a gel by a DNA molecule has a roughly logarithmic relationship with the m.w. of the DNA.

(f) Higher concentration of DNA may retard the movement of the DNA band. Higher concentration of salts in the buffer may also have the same effect.

(g) The RNA normally runs well in front of the DNA. Chromosomal DNA may be stuck in the well or move very slowly if present.

(h) Smear in the lane may indicate the degradation of the DNA by nucleases or maybe RNA. See also trouble-shooting

(i) To prepare low m.p agarose gel - first pour 50 ml (volume depends on gel apparatus used) of 2% of usual agarose (not LMP), let it set, put comb in (leave a small gap between gel surface and comb) and pour 100ml of ~1% LMP agarose. This method helps the low m.p agarose to set and provide a firm base for the handling of the gel.

(j) When loading DNA sample into the well of your gel, if the DNA sample is contaminated with ethanol or mineral oil, or if the gel is not completely submerged with buffer, the DNA solution may move out of the well. If this is a problem, use more loading buffer, or prepare the loading buffer in TAE instead of water, or use glycerol in the buffer and mix well, or make sure that your DNA sample is free of ethanol.

(k) Note that UV can damage your DNA. Although the UV transilluminator should be operating at a wavelength safer for your DNA, it is generally a good idea not to expose your DNA to UV light for too long. Overexposure to UV can result in ligation failure. It has been suggested that addition of 1-10mM cytodine or guanosine in gel can protect against UV damage.

(l) Electrophoresis of the gel can be done at constant current or constant voltage, however, it is safer to use constant voltage.

(m) You can destain the gel after staining (10-15 min. in water), but it is generally unncessary but would be useful if the band is faint.

METHODS OF EXTRACTING DNA FROM GEL

There are many methods for extracting DNA from gel apart from using commericial kits, some methods however yield cleaner DNA than others.

1. **Freeze squeeze**

   Excise band and place pieces of it in a home-made spin-column (a piece of filter placed in a PCR 0.5ml eppendorf, pierced the bottom of the PCR tube with a needle). Place The PCR tube at -20 °C for 15-20 mins or freeze in liquid N2. Then place the PCR tube into the 1.5 ml eppendorf and spin at top speed in a microfuge for 15 mins, collect DNA solution which can be EtOH ppt or use directly.

2. **Powdered silica**

   Melt excised bands at 70°C cool and add equal volume of TE-buffered phenol and mix. Spin for 5 min, remove aqueous phase, repeat and EtOh ppt. the DNA.

3. **Electroelution**

Cut a small trough just ahead of the DNA band, place DEAE-cellulose paper, or dialysis tubing, or affinity membrane, or just buffer containing 0.3 M NaOAc and 10% sucrose into the trough and run the gel at 150 volt for a minute or two until all the DNA has gone into the trough. Application of current will cause the DNA to migrate out of the agarose, but it will be trapped within the bag. When the DNA is out of the agarose, the flow of current is reversed for a few seconds to knock the DNA off of the side of the tubing. The buffer containing the DNA is then collected and the DNA precipitated with ethanol. Progress can be monitored using a transilluminator.

BUFFERS

There are a number of buffers used in this process, like

50x TAE (per litre)

- 242 g Tris base,
- 57.1 g glacial acetic acid,
- 100 ml 0.5 M EDTA
- pH 8.0

10x TBE (per litre)

- 108 g Tris base,
- 55 g boric acid,
- 40 ml 0.5 M EDTA,
- pH 8.0

20X TTE buffer (per litre)

- Tris Base 216 g,
- Taurine 72 g,
- EDTA disodium salt 4 g

6x gel loading buffer

- 0.25% Bromophenol blue,
- 0.25%Xylene cyanol FF,

- 15% Ficoll Type 4000,
- 120 mM EDTA

DETERMINATION OF DNA CONCENTRATION

To determine the concentration of DNA read the OD at 260nm in a spectrophotometer. An OD 260 of 1.0 give ~50 µg of dsDNA or ~37 µg of ssDNA. Check for purity of DNA by reading the OD 280 of the DNA solution as well. OD 260/OD 280 should give a value of 1.8 for pure DNA. Other values indicate the presence of contaminant such as proteins, RNA, and phenol.

HOW MUCH GEL SHOULD BE TAKEN?

Most agarose gels are made between 0.7% and 2%. A 0.7% gel will show good separation (resolution) of large DNA fragments (5–10kb) and a 2% gel will show good resolution for small fragments (0.2–1kb). Some people go as high as 3% for separating very tiny fragments but a vertical polyacrylamide gel is more appropriate in this case. Low percentage gels are very weak and may break when you try to lift them. High percentage gels are often brittle and do not set evenly. I usually make 1% gels.

USE OF ASMALL AND BIG GEL TANKS

Small 8x10cm gels (minigels) are very popular and give good photographs. Larger gels are used for applications such as Southern and Northern blotting. The volume of agarose required for a minigel is around 30–50mL, for a larger gel it may be 250mL. This method assumes you are making a mini-gel.

FOR ELECTROPHORESIS HOW MUCH AMOUNT OF DNA SHOULD BE LOADED ?

You may be preparing an analytical gel to just look at your DNA. Alternatively, you may be preparing a preparative gel to separate a DNA fragment before cutting it out of the gel for further treatment. Either way you want to be able to see the DNA bands under UV light in an ethidium-bromide-stained gel. Typically, a band is easily visible if it contains about 20ng of DNA.

Too much DNA should never be loaded onto a gel. The band appears to run fast (implying that it is smaller than it really is) and in extreme cases can mess up the electrical field for the other bands, making them appear the wrong size also. Too little DNA is only a problem in that you will not be able to see the smallest bands because they are too faint.

TYPE OF COMB THAT SHOULD BE USED?

This depends on the volume of DNA you are loading and the number of samples. Combs with many tiny teeth may hold 10µL. This is no good if you want to load 20µL of restriction digest plus 5µL of loading buffer. When deciding whether a comb has enough teeth, remember that you need to load at least one marker lane, preferably two.

PROCEDURE

- Making the gel (for a 1% gel, 50mL volume)
- Weigh out 0.5g of agarose into a 250mL conical flask. Add 50mL of 0.5xTBE, swirl to mix. It is good to use a large container, as long as it fits in the microwave, because the agarose boils over easily.

- Microwave for about 1 minute to dissolve the agarose. The agarose solution can boil over very easily so keep checking it. It is good to stop it after 45 seconds and give it a swirl. It can become superheated and NOT boil until you take it out whereupon it boils out all over you hands. So **wear gloves** and hold it at arms length. You can use a bunsen burner instead of a microwave - just remember to keep watching it.
- Leave it to cool on the bench for 5 minutes down to about 60°C (just too hot to keep holding in bare hands). If you had to boil it for a long time to dissolve the agarose then you may have lost some water to water-vapour. You can weigh the flask before and after heating and add in a little distilled water to make up this lost volume. While the agarose is cooling, prepare the gel tank ready, on a level surface.
- Add 1μL of ethidium bromide (10mg/mL) and swirl to mix. The reason for allowing the agarose to cool a little before this step is to minimise production of ethidium bromide vapour. Ethidium Bromide is mutagenic and should be handled with extreme caution. Dispose of the contaminated tip into a dedicated ethidium bromide waste container. 10mg/mL ethidium bromide solution is made up using tablets (to avoid weighing out powder) and is stored at 4°C in the dark with TOXIC labels on it.
- Pour the gel slowly into the tank. Push any bubbles away to the side using a disposable tip. Insert the comb and double check that it is correctly positioned. The benefit of pouring slowly is that most bubbles stay up in the flask. Rinse out the flask immediately.
- Leave to set for at least 30 minutes, preferably 1 hour, with the lid on if possible. The gel may look set much sooner but running DNA into a gel too soon can give terrible-looking results with smeary diffuse bands.
- Pour 0.5x TBE buffer into the gel tank to submerge the gel to 2–5mm depth. This is the running buffer. You must use the same buffer at this stage as you used to make the gel. ie. If you used 0.6xTBE in the gel then use 0.6xTBE for the running buffer. Remember to remove the metal gel-formers if your gel tank uses them.

PREPARING THE SAMPLES

Transfer an appropriate amount of each sample to a fresh microfuge tube. Write in your lab-book the physical order of the tubes so you can identify the lanes on the gel photograph.

- Add an appropriate amount of loading buffer into each tube and leave the tip in the tube. Add 0.2 volumes of loading buffer, eg. 2μL into a 10μL sample. The tip will be used again to load the gel.
- Load the first well with marker. Avoid using the end wells if possible. For example, If you have 12 samples and 2 markers then you will use 14 lanes in total. If your comb formed 18 wells then you will not be using 4 wells. It is best to not use the outer wells because they are the most likely to run aberrantly.
- Continue loading the samples and finish of with a final lane of marker. load gels from right to left with the wells facing you. This is because gels are published, by convention, as if the wells were at the top and the DNA had run down the page. If this seems confusing then you can load left to right with the wells facing away from you.
- Close the gel tank, switch on the power-source and run the gel at 5V/cm. For example, if the electrodes are 10cm apart then run the gel at 50V. It is fine to run the gel slower than this but do not run any faster. Above 5V/cm the agarose may heat up and begin to melt with disastrous effects on your gel's resolution. Some people run the gel slowly at first (eg. 2V/cm for 10 minutes) to allow

the DNA to move into the gel slowly and evenly, and then speed up the gel later. This may give better resolution. It is OK to run gels overnight at very low voltages, eg. 0.25–0.5V/cm, if you want to go home at 11 O'clock already.

- Check that a current is flowing. The best way is to look at the electrodes and check that they are evolving gas (ie. bubbles).
- Monitor the progress of the gel by reference to the marker dye. Stop the gel when the bromophenol blue has run 3/4 the length of the gel.
- Switch off and unplug the gel tank and carry the gel (in its holder if possible) to the dark-room to look at on the UV light-box. Some gel holders are not UV transparent so you have to carefully place the gel onto the glass surface of the light-box. UV is **carcinogenic** and must not be allowed to shine on naked skin or eyes. So wear face protection, gloves and long sleeves.

LOADING BUFFERS

The loading buffer gives colour and density to the sample to make it easy to load into the wells. Also, the dyes are negatively charged in neutral buffers and thus move in the same direction as the DNA during electrophoresis. This allows you to monitor the progress of the gel. The most common dyes are bromophenol blue (Sigma B8026) and xylene cyanol (Sigma X4126). Density is provided by glycerol or sucrose.

COMPOSITION

- 25mg bromophenol blue or xylene cyanol
- 4g sucrose
- H_2O to 10mL

The exact amount of dye is not important Store at 4°C to avoid mould growing in the sucrose. 10mL of loading buffer will last for years. Bromophenol blue migrates at a rate equivalent to 200–400bp DNA. If you want to see fragments anywhere near this size (ie. anything smaller than 600bp) then use the other dye because the bromophenol blue will obscure the visibility of the small fragments. Xylene cyanol migrates at approximately 4kb equivalence. So do not use this if you want to visualise fragments of 4kb.

TBE

TBE stands for Tris Borate EDTA.

People also use TAE (Tris Acetate EDTA). Make up a 10x stock using cheap reagents. Do not use expensive 'analytical grade' reageants. Cheap Tris base and boric acid can be bought in bulk.

COMPOSITION FOR 2L OF 10XTBE

- 218g Tris base
- 110g Boric acid
- 9.3g EDTA

Dissolve the ingredients in 1.9L of distilled water. pH to about 8.3 using NaOH and make up to 2L.

51

RNA GEL ELECTROPHORESIS

FORMALDEHYDE-AGAROSE GELS METHOD

Reagents

- 20 × MOPS buffer 400 mM 3-(N-morpholino)
- Propanesulphonic acid,
- 160 mM sodium acetate,
- 20 mM EDTA, pH 7.0. Sterilise by autoclaving
- Ethidium bromide 10 mg / ml in sterile water
- Agarose
- Formaldehyde
- RNA loading buffer 1 x MOPS buffer (ie, 20 mM MOPS, 8 mM sodium acetate and 1 mM EDTA, pH 7.0), 7% (w/v) formaldehyde pH 4.0, 5% (v/v)
- Sterile glycerol, 50% (v/v)
- Deionised formamide
- 0.025% (v/v) of saturated aqueous bromophenol blue solution
- RNA samples

EQUIPMENT

- Gel former
- Gel combs
- Electrophoresis power supply

METHODS

Preparation of agarose gel

1 Clean the gel-former and comb with distilled water, then 70% ethanol. Seal the edges of the gel-former with tape. Check that the comb sits approximately 1 mm above the gel-former when in situ 2 Make up enough buffer for both the electrophoresis tank and the gel to avoid any differences in ionic strength between them 3 For 100 µl of a 1.3% gel : melt 1.3 g of agarose in 50 µl water. Add to 20 µl of formaldehyde 6, 5 µl of 20 x MOPS and 5 µl of ethidium bromide solution in a measuring cylinder. Make the volume up to 100 µl with MilliQ water 4 Swirl to mix and then pour the gel carefully, checking for air bubbles under or between the teeth7. The final gel should be between 3 mm and 5 mm thick. Allow it to set at room temperature for 30 - 45 minutes 8, 9, 5 When fully set, pour enough buffer to cover the gel

surface by ~ 1 mm and allow to stand for a couple of minutes 10. Remove the comb carefully to avoid tearing the bottom of the wells (and subsequently losing the sample), and place the gel in the electrophoresis tank. Cover to a depth of ~ 1 mm

PREPARE RNA SAMPLES

6 Prepare the RNA to be analysed. The loading buffer can be diluted by up to 2 fold although generally 5 μl of aqueous RNA is added to 15 μl of RNA loading buffer11 7 Denature the RNA samples in loading buffer by heating to 70oC for 5 minutes 8 Chill on ice before loading the gel 12, 13, 14,

Electrophoresis through agarose or polyacrylamide gels is the standard way to separate, identify and purify nucleic acid fragments. The location of the nucleic acid within in the gel can be determined by using the fluorescent intercalating dye ethidium bromide.

Agarose gels have a smaller resolving power than polyacrylamide gels but a greater range of separation - from 200 bp to >50 kb using standard gels and electrophoresis equipment. RNAs up to 10 000 kb can be separated in agarose gels using pulsed field gel electrophoresis.

Polyacrylamide gels have enough resolving power to separate fragments differing by only one base pair in size, but their range is ~ 5 to 1000 bp. They are much more difficult to handle than agarose gels.

FORMALDEHYDE-AGAROSE GEL ELECTROPHORESIS

Agarose is a polysaccharide obtained from seaweed. There are frequently contaminants - other polysaccharides, salts and proteins - and different batches as well as different manufacturers brands vary in the level of contaminants and hence in the performance of the agarose. Agarose can be chemically modified to gel and melt at lower temperatures by the addition of hydroxyethyl groups into the polysaccharide chain.

Agarose gels are cast by completely melting the agarose in the desired buffer and then pouring into a mould to harden. RNA is negatively charged at neutral pH and when an electric field is applied, it migrates towards the anode.

RNA retains much of its secondary structure during electrophoresis unless it is first denatured. The addition of formaldehyde to the agarose gel maintains the RNA in its linear (denatured) form.

THE RATE OF MIGRATION IS DETERMINED BY

• Molecular Size of RNA

Linear RNA becomes orientated in an electric field in an 'end-on' position and migrates through the matrix of the gel at a rate which is inversely proportional to the log10 of the number of base pairs. Larger molecules migrate more slowly because of greater frictional drag as they try to pass through the gel matrix

• Agarose Concentration

Linear RNA of a given size migrates through agarose of different concentrations at different rates given by log m = logmo -KrT, where m is the electrophoretic mobility of the RNA, mo is the free electrophoretic mobility of the RNA, Kr is the retardation coefficient and T is the gel concentration

• Conformation of the RNA

RNA molecules which fully or partially retain their secondary structure, migrate at different rates to fully denatured RNAs with the same molecular mass

% of agarose (w/v) in gel	Efficient range of separation of linear RNA molecules - bp
0.3	5 000–60000
0.6	1 000–20000
0.7	800–10000
0.9	500–7000
1.2	400–6000
1.5	200–3000
2.0	100–2000

• Applied voltage

At low voltage, the rate of migration of RNA is proportional to the voltage applied. As the voltage is increased, the mobility of larger molecules increases differentially. the effective range of separation therefore decreases with increasing voltage. For maximum resolution, run agarose gels at no more than 5V / cm (measured between the electrodes, not the gel length).

• Base composition of the RNA and temperature of the gel

For agarose, neither of these parameters significantly affect the mobilities of RNA

• Presence of intercalating dyes

Ethidium bromide reduces the electrophoretic mobility of linear RNA by about 15%. EtBr has greater affinity for double than single-stranded nucleic acids.

• Composition of the electrophoresis buffer

In the absence of ions, RNA migrates very slowly if at all. If a 10 x buffer is used by mistake, electrical conductance is very efficient, the current generates a lot of heat and melt down happens.

ELECTROPHORESIS OF RNA SAMPLE

9 Load the RNA samples carefully into the slots. Connect the electrophoresis tank to a constant voltage power supply - RNA will run from black to red ie cathode to anode : make sure leads are on the right way round ! Run at 1 - 5V / cm (measured as the distance between the electrodes) until the BPB dye front has migrated the appropriate distance. Check after the BPB has run 50% of the way down the gel. The ethidium bromide will migrate the opposite way to the RNA and long electrophoresis will remove much of the ethidium from the gel15

PROCEDURE

Preparation of RNA sample for electrophoresis

- Mix 1 volume of the 2X RNA Loading Dye Solution and 1 volume of RNA sample.
- Heat at 70°C for 10min.
- Chill on ice for 3 minutes and load onto the gel.

Note

RNA samples prepared as described above are not suitable for glyoxal/DMSO agarose gel electrophoresis. To prepare RNA samples for glyoxal/DMSO agarose gels, please refer to the recommendations for glyoxal/DMSO agarose electrophoresis.

NON-DENATURING AGAROSE GEL ELECTROPHORESIS

Dilute 50X TAE buffer or 10X TBE buffer to a 1X concentration immediately before use.

Note

Use TBE buffer for analysis RNA bands smaller than 1500b. For larger RNA, use TAE buffer.

Prepare 1% TopVision™ LE GQ Agarose gel in 1X TAE or TBE according to the recommended protocol.

Note

For optimal results, add ethidium bromide to both the agarose gel and electrophoresis buffer at a final concentration of 0.5µg/ml.

Alternatively, the gel can be stained after electrophoresis by immersing it into a 0.5µg/ml ethidium bromide solution for 20min, or by any other RNA staining technique. Place the gel into an electrophoresis apparatus containing the appropriate buffer. Heat the RNA samples and ladder at 70°C for 10min, and then chill on ice for 3 minutes. Load onto the gel. Run electrophoresis at 5V/cm until the bromophenol blue runs approximately two-thirds of the way down the gel.

DENATURING FORMALDEHYDE GELS IN MOPS BUFFER

Freshly prepare 10X MOPS buffer:

- 0.4M MOPS (pH 7.0),
- 0.1M Sodium acetate,
- 0.01M EDTA (pH 8.0).

PREPARE 1% AGAROSE GEL AS FOLLOWS

- stir 1g of agarose powder in 72ml of deionized water
- melt the agarose, and then add 10ml of 10X MOPS buffer and mix
- when the agarose solution cools to 60°C, add 18ml of fresh formaldehyde (37%) in a fume hood and mix thoroughly
- pour the gel.
- Place the gel into an electrophoresis apparatus containing 1X MOPS buffer.

Heat the RNA samples and ladder at 70°C for 10min, and then chill on ice for 3 minutes. Load onto the gel.

Note

There is no need to stain the gel as ethidium bromide present in 2X RNA Loading Dye Solution is sufficient for visualization under UV light.

Sample loading buffer

Deionized formamide	-	**750** μl
10 × MOPS-Acetate	-	150 μl
Formaldehyde AR grade (37%)	-	240 μl
Sterile ddH$_2$O (DEPC treated & autoclaved)	-	75 μl
Glycerol (autoclaved)	-	100 μl
3% Bromophenol blue (sterile)	-	25 μl

Preparation of 1.2% agarose gel for RNA electrophoresis

Sterile ddH$_2$O	-	84.9 μl
Agarose	-	1.2 g, boil to dissolve and then add
10 X MOPs Acetate	-	10.0 μl
37% Formaldehyde	-	5.1 μl (0.66 M)
Sterile ddH$_2$O to	-	100.0 μl

Preparation of RNA sample for loading

RNA (5 to 10 mg)	-	5.0 μl
Sample loading buffer	-	25 μl

Denaturing glyoxal/DMSO gels in sodium phosphate buffer

- Prepare thick 1.0% Agarose gel in 0.01M sodium phosphate buffer, pH 7.0.
- Place the gel into an electrophoresis apparatus containing 0.01M sodium phosphate buffer, pH 7.0.
- Prepare for loading 25μl aliquots of the ladder/samples by adding:
- Glyoxal (40% solution) 4.5μl
- DMSO 12.5μl
- 0.1M sodium phosphate buffer, pH 7.0 2.5μl
- Mix and add RNA 3μl
- 2X RNA Loading Dye Solution 1μl
- DEPC-treated Water to 25μl

INCUBATE FOR 1 HOUR AT 50°C AND THEN COOL DOWN TO ROOM TEMPERATURE

- Load the samples on a gel.
- Run electrophoresis at 5V/cm until the bromophenol blue runs approximately two-thirds of the way down the gel
- Stain the gel in ethidium bromide solution (final concentration 0.5μg/ml) in 0.5M ammonium acetate for 15-30min
- Wash the gel in fresh 0.5M ammonium acetate solution for 15-30min.

DENATURING POLYACRYLAMIDE/UREA GELS IN TBE BUFFER

- Prepare 20ml of a 5% polyacrylamide gel containing 7M urea by adding:
- 47.5% acrylamide:
- 2.5% bis-acrylamide solution 2ml
- 10M urea 14ml
- 10X TBE buffer 2ml
- 10% freshly prepared ammonium persulfate 0.2ml
- deionized water 1.8ml

MIX AND ADD 10µL TEMED

- Mix again and pour the gel carefully avoiding the formation of air bubbles.
- Insert the comb into the acrylamide and allow the gel to polymerize for at least 1 hour.
- Fill the electrophoresis apparatus with 1X TBE buffer
- Heat the RNA samples (premixed with equal volume of 2X RNA Loading Dye Solution) at 70°C for 10min, and chill on ice for 3 minutes. Load onto the gel.
- Run electrophoresis at 8V/cm for about 1 hour.
- Soak the gel for about 15 minutes in 1X TBE to remove urea prior to staining.
- Stain the gel in 0.5 µg/ml ethidium bromide in 1X TBE solution for 15min.

52

POLYACRYLAMIDE GEL ELECTROPHORESIS

Monomeric acrylamide (which is neurotoxic) is polymerised in the presence of free radicals to form polyacrylamide. The free radicles are provided by ammonium persulphate and stabilised by TEMED (N'N'N'N'-tetramethylethylene-diamine). The chains of polyacrylamide are cross-linked by the addition of methylenebisacrylamide to form a gel whose porosity is determined by the length of chains and the degree of crosslinking. The chain length is proportional to the acrylamide concentration : usually between 3.5 and 20%. Cross-linking BIS-acrylamide is usually added at a ratio of 2g BIS 38g acrylamide.

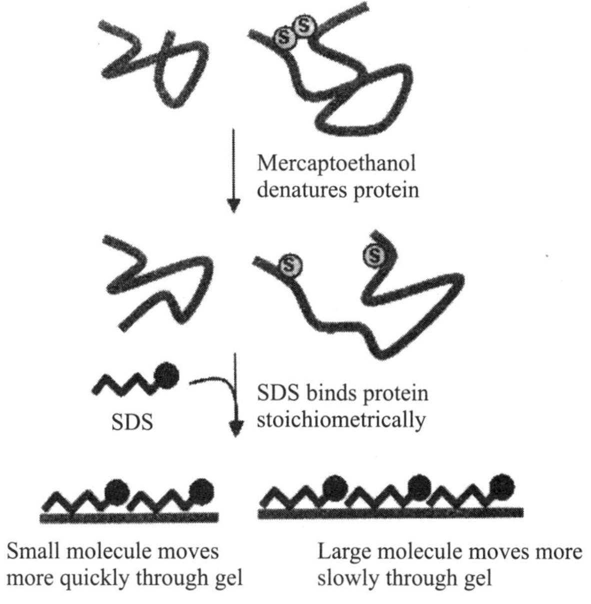

Fig. 51 SDS PAGE electrophoresis

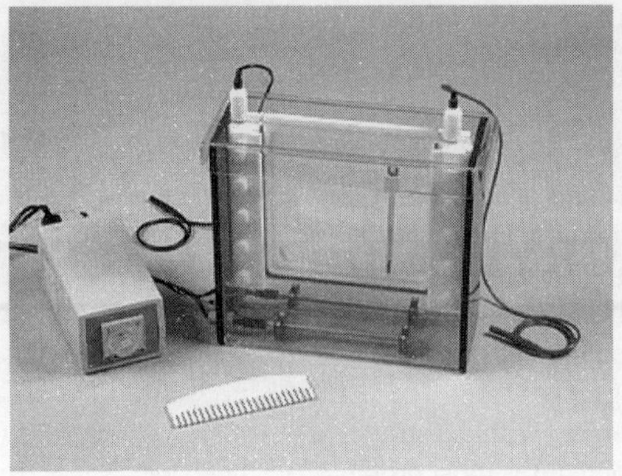

Fig. 52 Vertical SDS PAGE Electrophoresis Unit

% age acrylamide (w/v) with BIS at 1:20	Effective range of separation - bp	Size of RNA co-migrating with Xylene Cyanol	Size of RNA co-migrating with Bromophenol Blue
3.5	1 000 - 2 000	460	100
5.0	80 - 500	260	65
8.0	60 - 400s	160	45
12.0	40 - 200	70	20
15.0	25 - 150	60	15
20.0	6 - 100	45	12

Polyacrylamide gels are poured between two glass plates held apart by spacers of 0.4 - 1.0 mm and sealed with tape. Most of the acrylamide solution is shielded from oxygen so that inhibition of polymerisation is confined to the very top portion of the gel. The length of the gel can vary between 10 cm and 100 cm depending on the separation required. They are always run vertically with 0.5 / 1 x TBE as a buffer.

THREE MAIN ADVANTAGES OVER AGAROSE GELS

1. Resolving power is such that RNA molecules differing in length by 1 base in 500 ie 0.2% can be effectively separated.
2. They can hold up to 10 mg of RNA per slot (20 mg of RNA) without loss of resolution, a much larger amount than agarose gels
3. The recovered RNA or RNA is extremely pure.

TWO TYPES OF POLYACRYLAMIDE GEL IN GENERAL USE

• Non-denaturing gels

These are run at low voltages - 8V/cm - and 1 x TBE to prevent denaturation of small fragments of RNA by the heat generated in the gel during electrophoresis. The rate of migration is approximately inversely proportional to log10 of their size. However, the base sequence composition can alter the electrophoretic mobility of RNAs such that two RNAs of the same size may show up to a 10% difference in electrophoretic mobility

• Denaturing gels

These gels are polymerised with a denaturant that suppresses base pairing in nucleic acids - this is usually urea but can be formamide. Denatured RNA migrates through the gel at a rate which is almost completely independent of its composition or sequence. These gels are used for the analysis of sequencing reactions, RNase protection assays and purification of radiolabelled RNA and RNA probes.

A common technique used to simplify the interpretation of electrophoretic runs in a single gel is to run the gel under denaturing conditions. The denaturant of choice is usually sodium dodecyl sulfate (SDS), which is an ionic detergent with the structure $CH_3(CH_2)_{10}CH_2OSO_3^-$ (a single chain amphiphile). This detergent binds to and denatures most proteins, with about 1.4 g SDS binding/g of protein (about 1 SDS/2 amino acids). Since there is 1 negative charge/SDS, the binding of SDS masks any of the charges on the protein, and gives all proteins an overall large negative charge. *Additionally*, SDS-proteins complexes have been shown to generally have a elongated cylindrical-like shape. Since the amount of SDS bound per unit mass of protein is constant, the overall charge density on all proteins is similar, so the electrophoretic mobility is only determined by sieving effects. SDS also eliminates shape differences in the proteins as a variable which determines sieving, since all proteins have the same general rod-like shape. (The use of SDS is analogous to the use of 8M urea in the gel chromatographic separation of proteins to determine molecular weights). Mobility becomes only a a function of the molecular weight of the protein, and not shape. The molecular weight of an unknown protein can be determined by comparing the protein's position on an SDS polyacrylamide gel with a series of known molecular weight standards from which a linear plot of the ln Mr vs R_f can be used to calculate unknown molecular weights. This is similar to the analysis in gel chromatography, where ln Mr is a linear function of K_{avg}, the distribution coefficient, when the gel is run under denaturing conditions. However, some proteins run anomalously on such gels (due to incomplete or excess binding SDS), so alternative techniques of molecular weight determination should be used in conjunction with this technique.

Proteins are usually heated in SDS to 100°C for 3 minutes, in the presence of a reducing agent such as b-mercaptoethanol, to completely denature the protein to a rod-shaped protein. Apparent molecular weight can be obtained under non-reducing conditions (without b-ME), but these should be considered just estimates. Running proteins both in the presence and absence of the reducing agent can provide important information on the subunit structure of a protein. A multimeric protein whose subunits are held together by disulfide bonds can be resolved into its individual components when the reducing agent is added. If the subunits are held together by noncovalent intermolecular attractions, the proteins will run identically under the denaturing conditions (SDS), which will eliminate subunit interactions, in the presence or absence of b-ME. To determine the subunit composition of a protein held together by noncovalent interactions, the electrophoresis should be performed in the absence of denaturing agents.

TECHNIQUES

Polymerizing and pouring the gel

Electrophoresis is performed in a porous, yet solid medium, to eliminate any problems associated with convection currents. Such media are formed from the polymerization of a liquid solution of agarose (used mostly for electrophoresis of DNA fragments and very large proteins) or acrylamide. Polymerization of acrylamide is initiated by the additions of ammonium persulfate in the presence of tetramethylenediamine (TEMED), along with a dimer of acrylamide (N,N'-methylene-bis(acrylamide) connected covalently between the amide nitrogens of the acrylamides by a methylene group. The structures of these compounds is shown below:

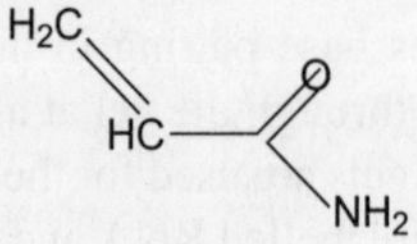

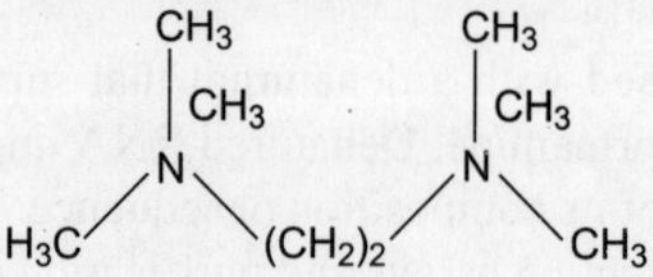

acrylamide   retramethylethylenediamine (TEMED)

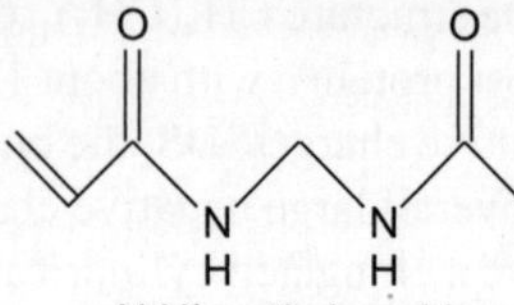

N,N'-methylene-bis-acrylamide

The free radical polymerization of the acrylamide is initiated on the addition of ammonium persulfate, which on dissolving in water, forms free radicals, as shown below:

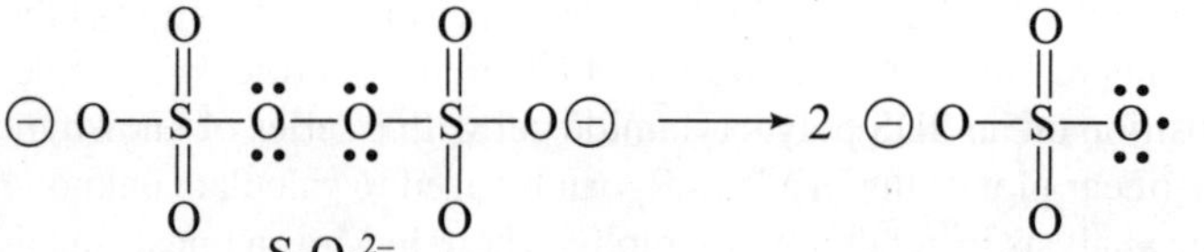

$$S_2O_8^{2-} \text{ (Persulfate)} \longrightarrow 2\ SO_4^{-}$$

The TEMED, through its ability to exist as a free radical, acts as an additional catalyst for the polymerization. A rigid gel is only formed, however, when N,N'-methylene-bis(acrylamide is added to the mixture during the polymerization, which cross-links adjacent acrylamide polymers.

The amount of bis added during polymerization controls the degree of cross-linking, and hence the pore size of the polymerized gel. The effect of pore size is opposite to that in gel chromatography. In both cases, large proteins have a difficult time entering the pore. In gel chromatography, large proteins partition preferentially into the mobile liquid phase (the void volume) and are eluted most quickly from the column. In electrophoresis, large proteins, which can not readily enter the pores in the gel, are not as easily transported by the electric field through the gel, and elute most slowly. Pore size can not be controlled as accurately as in the manufacture of gel chromatography resins.

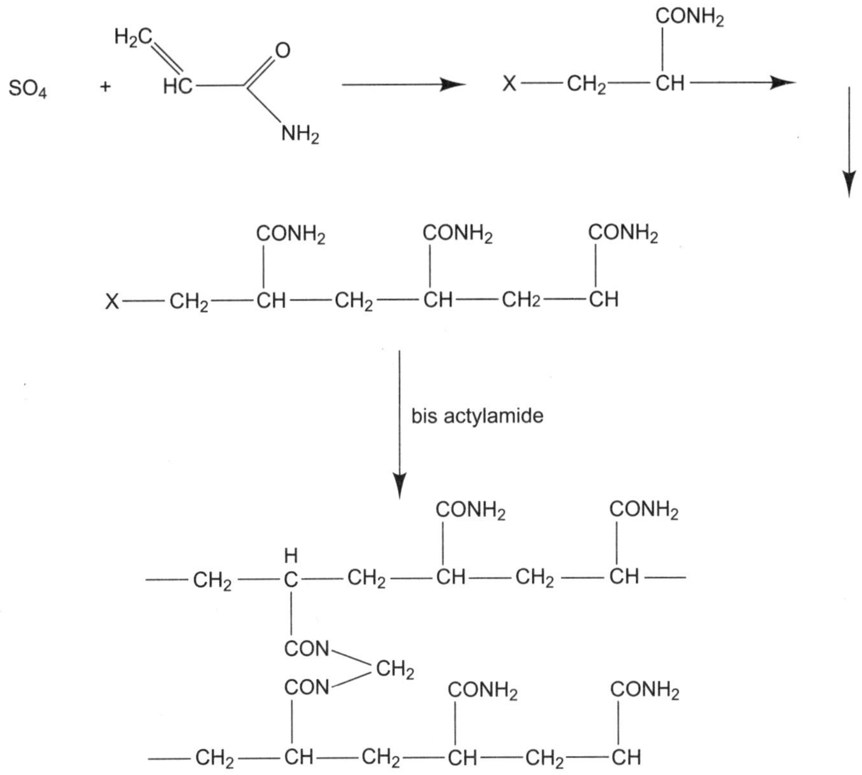
SO_4 + H_2C HC O NH_2
X — CH_2 — CH — $CONH_2$
X — CH_2 — CH — CH_2 — CH — CH2 — CH
$CONH_2$ $CONH_2$ $CONH_2$
bis actylamide
— CH_2 — C — CH_2 — CH — CH_2 — CH —
H
CON
CON
CH_2
$CONH_2$ $CONH_2$
$CONH_2$ $CONH_2$
— CH_2 — CH — CH_2 — CH — CH_2 — CH

How do proteins migrate through the gel? A viscous protein solution is layered on the top of the gel in a small well molded into the gel during the polymerization process. The bottom and top parts of the gel are inserted into reservoirs containing a buffered solution and the appropriate electrode. The electric field is applied and the proteins migrate through the hydrated gel. The nature of the buffer solution in the reservoir and in the polymerized gel is important. The components of the buffer must not bind to the proteins to be separated. Additionally, the pH of the medium must be such that the proteins have the appropriate charge, so they will migrate in the expected direction.

DISCONTINUOUS GEL ELECTROPHORESIS

There are many variations of electrophoresis commonly used. Gels can be polymerized in tubes, or slabs, and in the presence or absence of denaturing agents. Additionally, a given slab might consist of two separate slabs polymerized one on top of each other, each with a different acrylamide concentration and pH. This type is called discontinuous pH gel electrophoresis, or disc-electrophoresis.

The stacking gel is a low concentration acrylamide (2-4%) polymerized in a Tris HCl buffer solution (pH 6.5) two pH units below that used in the running gel and the bottom reservoir (Tris HCl buffer, pH 8.7). The lower or running gel concentration varies from 7-15% acrylamide, depending on the molecular weight of the proteins to be separated. The upper buffer reservoir contains Tris buffered with a weak acid such as glycine (pKa2 = 9.6) to the same pH as the running gel.

Polyacrlymaide concentration and porosity of gels

% of polyacrylamide	Average pore diameter (^{o}A)
3	44
5	36
7.5	30
10	26
15	22
20	18
25	15
30	13
35	12

PREPARATION OF SOLUTIONS

(1) Acrylamide-bisacrylamide solution:

Acrylamide	-	29.2 g
Bis acrylamide	-	00.8 g
dH$_2$O to	-	100 ml

Stir the solution for 15 min with a small amount of activated charcoal and later filter through nitrocellulose membrane filter, stored at 4° C in a brown bottle. (Never store the solution for a long time, as acrylamide becomes acrylic acid).

(2) 4x Separating gel buffer (1.5 M Tris, pH 8.8):

Dissolve 18.15 g of 1.5 M Tris in 80 ml dH$_2$O, adjust the pH to 8.8 with conc. HCl and make up the volume to 100 ml.

(3) 4x Stacking gel buffer (0.5 M Tris, pH 6.8)

Dissolve 3.0 g of 0.5M Tris in 30 ml dH$_2$O, adjust the pH to 6.8 with conc. HCl and make up the volume to 50 ml.

(4) 10% SDS

Dissolve 10g of SDS in 100ml of dH$_2$O

(5) Initiator: (10% APS)

Ammonium persulphate	-	0.1 g
dH$_2$O	-	1.0 ml

Dissolve 0.1g of APS in 1.0ml of dH$_2$O and store in ice.

(6) Catalyst

TEMED is supplied in brown bottle and stored at 4° C.

(7) Separating gel overlaying solution

n – or iso-butanol	-	50 ml
dH_2O	-	50 ml

Mix 50ml of n – or iso-butanol with 50ml of dH2O. Separate the water saturated butanol (upper layer) after shaking the mixture vigorously and then allowing it to settle.

(8) Bromophenol blue (0.1%)

Dissolve 5 mg bromophenol blue in 5 ml water.

(9) 2 x sample buffer

Tris (0.5 M, pH 6.8)	-	2.5 ml (4 x Stacking gel buffer)
SDS (10%)	-	4.0 ml
Glycerol (100%)	-	2.0 ml
â-mercaptoethanol	-	0.8 ml (or IM DDT – 0.5 ml)
Bromophenol blue (0.1%)	-	300 µl
dH_2O (400 µl) to	-	10.0 ml

(10) Tank buffer

Tris	-	6.05 g
Glycine	-	28.80 g
10% SDS	-	10.0 ml or (1.0 g)
dH_2O	-	1000 ml

(11) Staining solution

Coomassie blue (R-250)	-	0.3 g
Methanol (AR)	-	80 ml
Glacial acetic acid	-	20 ml
dH_2O	-	100 ml

(12) Destaining solution

Acedic acid	-	100 ml
Methanol	-	300 ml
dH_2O	-	1.0 litre

(13) Gel storing solution

Acetic acid	-	15 ml
dH_2O	-	200 ml

(14) Standard proteins (2 ìg/ì1)

Phosphorylase a (97.4 kDa)	-	2 mg/ml
BSA (66 kDa)	-	2 mg/ml
Egg albumin (45 kDa)	-	2 mg/ml

Carbonic anhydrase (29 kDa)   -   2 mg/ml
Lysozyme (14.4 kDa)   -   2 mg/ml

(15) Preparation of Separating gel – 10% (10 ml is required for baby gels):

dH$_2$O   -   4.00 ml
30% acrylamide:bis acrylamide   -   3.33 ml
4 x Tris (pH 8.8)   -   2.50 ml
TEMED   -   5 µl
10% SDS   -   100 µl
10% APS (100 mg/ml)   -   50 µl
10.00 ml

(16) Preparation of Stacking gel – 4% (5 ml is required for baby gels):

dH$_2$O   -   3.00 ml
30% acrylamide: bis   -   0.67 ml
4 x Tris (pH 6.8)   -   1.25 ml
TEMED   -   2.5 µl
10% SDS   -   25 µl
10% APS   -   50 µl

SDS-PAGE OF PROTEIN SOLUTIONS

(1) Add 100 ml of 0.15% sodium deoxycholate (15mg/10 ml dd H$_2$O) to 1 ml of the dilute protein solution in an Eppendorf tube.
(2) Leave it for 10 min at room temperature.
(3) Add and mix gently 150 ml 50% cold TCA.
(4) Incubate the appendorf in ice for 15 min.
(5) Centrifuge at 5000rpm for 10 min.
(6) Decant the supernatant and leave the tube inverted on a tissue paper.
(7) Dissolve the pellet in 10 ml of 0.25 N NaOH/KOH by gently tapping the appendorf.
(8) Add 10 ml of 2X loading dye and see that the solution turns blue, in case it does not add 2 ml 0.25 N NaOH until the solution becomes blue.
(9) Boil the solution for 3 min in water bath
(10) Centrifuge again for 2 min and load.

SILVER NITRATE STAINING

The principle of silver nitrate staining of proteins is based on the reduction of silver ions to its metallic form. (Silver ions in alkaline conditions bind with proteins through å-amino group of lysine and sulphur groups of cysteine and methionine residues). When the complexed silver ions are reduced in the presence of formaldehyde, they become metallic silver, which is seen on the gel as bands. Staining of proteins with silver nitrate after SDS-PAGE is quicker and ~ 10 times more sensitive than Coomassie brilliant blue staining and that is the reason it is widely used in research laboratories these days.

The procedure is as follows:

PREPARATION OF SOLUTIONS

1. Mix 50 ml water and 50 ml methanol (Fixing solution-1)
2. Mix 50 ml water and 35 ml methanol and 15 ml 37% formaldehyde (Fixing solution-2)
3. Sodium thiosulphate 0.22% (20 mg in 100 ml)- dilute 1:10 from 0.2% stock.
4. Dissolve 100 mg silver nitrate in 100 ml water (0.1% Staining solution)
5. Dissolve 3 g K_2CO_3 OR Na_2CO_3 in 100 ml water and add 2 ml of 0.02% thiosulphate solution and 0.05 ml 37% formaldehyde (freshly prepared) (Developing solution)
6. Dissolve 5 g citric acid in 100 ml water (Stopping solution)
7. Mix 10 ml glacial acetic acid 10 ml methanol and 80 ml water (Gel storing solution)

(For staining small amount of gels the volume of the above solutions can be reduced to half)

PROCEDURE

1. Transfer the gel to the Fixing solution-1 after electrophoresis and incubate for 30 min. (The gel can be stored for a long time in this solution) followed by dipping the gel in Fixing solution-2 for 10 min.
2. To remove excess methanol and formaldehyde Wash the gel twice with ~50 ml water for 5 min each time
3. Transfer the gel in 0.02% thiosulphate solution for one min and rinse with water.
4. Immerse the gel in silver nitrate solution for 15 min and later rinse thoroughly with distilled water to remove excess silver nitrate.
5. Incubated the gel in 100 ml of developing solution consisting of 3% K_2CO_3 solution containing 2.0 ml of 0.2% thiosulphate solution 0.05 ml formaldehyde until you see the bands.
6. Stop the reaction immediately by adding 100 ml of the citric acid solution.
7. The gel can be stored in 10% acetic acid methanol solution. (For long term storage 10% glycerol can be included).

53

WESTERN BLOT ANALYSIS

AIM

To perform western blot analysis

PRINCIPLE

Identification of a specific protein in a complex mixture of protein or Ab to a given protein can be accomplished by western blot. In this technique protein is electrophoretically separated on a polyacrylamide slab gel. The protein bands are transferred into a nitrocellulose membrane by electrophoresis or by diffusion such that the membrane gets a replica of the gel containing protein bands. Flooding the membrane with primary Ab identifies the individual proteins. After this the Ag – Ab can be detected by adding a secondary Ab conjugated with enzyme. The band is visualised by adding substrate, which being a Chromogen on conversion to product gives colour. The intensity of colour shows the quantity of Ag – Ab complex formed.

Sufficiently separated proteins in an SDS-PAGE can be transferred to a solid membrane for WB analysis. For this procedure, an electric current is applied to the gel so that the separated proteins transfer through the gel and onto the membrane in the same pattern as they separate on the SDS-PAGE. All sites on the membrane, which do not contain blotted protein from the gel, can then be non-specifically "blocked" so that antibody (serum) will not non-specifically bind to them, causing a false positive result.

To detect the antigen blotted on the membrane, a primary antibody (serum) is added at an appropriate dilution and incubated with the membrane. If there are any antibodies present which are directed against one or more of the blotted antigens, those antibodies will bind to the protein(s) while other antibodies will be washed away at the end of the incubation. In order to detect the antibodies, which have bound, anti-immunoglobulin antibodies coupled to a reporter group such as the enzyme alkaline phosphatase are added (e.g. Goat anti-human IgG- alkaline phosphatase). This anti-Ig-enzyme is commonly called a "second antibody" or "conjugate". Finally after excess second antibody is washed free of the blot, a substrate is added which will precipitate upon reaction with the conjugate resulting in a visible band where the primary antibody bound to the protein.

Enzymes and substrates for western blot analysis: A number of enzymes and substrates have been employed for western blot developing, like alkaline phosphatase, Horseradish Peroxidase, NBT, BCIP.

    1. Nitroblue Tetrazolium (NBT):   prepared in buffer

NBT	=	3.0mg
Buffer : Tris-HCl	=	0.1M (pH 8.8)

NaCl	=	0.1M
$MgCl_2$	=	0.005M

2. 5-Bromo-4-Chloro Indolyl Phosphate (BCIP):

BCIP	=	1.5mg
Dimethyl Sulphoxide	=	10µl
2M Tris-HCl (pH9.8)	=	0.25ml

Mix solution 1 and 2 and make up to 10ml

3. Horseradish Peroxidase (HRP):
   30mg of 4-Chloro-1-Naphthol in 10ml Methanol
   C. 30µl of H_2O_2 in 40ml TBS

Mix solution A and B at room temperature.

Mechanism of Transfer of protein from gel to membrane: Transfer of protein bands from an Acrylamide gel onto a membrane by diffusion by the capillary action of buffer is called 'Capillary Blotting'. For this a number of solutions are required.

1. Urea Buffer:

Sodium Chloride	=	0.29g
EDTA	=	0.07g
Tris	=	0.12g
Urea	=	24.00g

Adjust the pH to 7.0 and make the volume to 100ml with distilled water

2. Transfer Buffer:

Sodium Chloride	=	2.9g
EDTA	=	0.7g
Tris	=	1.2g

3. Phosphate Buffer saline (PBS- pH7.2))

Sodium Chloride	=	8.00g
Potassium chloride	=	0.2g
Disodium Hydrogen Phosphate	=	1.15g
Potassium dihydrogen Phosphate	=	0.20g
Distilled water	=	**1 litre**

54

CHROMATOGRAPHY

PRINCIPLE

The physical method of separation of simple and complex mixture of biological samples has become important in all fields of biological research. The term chromatography comes from the earlier times when the technique was used for the separation of colored plants pigments. Chromatography is a technique for separation of closely related groups of compounds. The separation is brought about by differential migration along a porous medium and the migration is caused by the flow of solvent. Chromatography provides one of the beast answers in this field. The different types of chromatography in common use are paper, thin-layer (TLC), column, and gas-liquid (GC, VPC, GLC). All of the chromatographic procedures involve the interaction of a mobile phase (either a gas or a liquid) and a stationary phase (either a liquid or a solid). Within limits chromatography can be divided into two types:

- Partition and
- Adsorption chromatography.

The four basic categories are:

TYPE	MOBILE	STATIONARY	PRINCIPLE OF SEPARATION	USES
column	liquid	solid	adsorption	Preparative scale separations
Thin-layer	liquid	solid	adsorption	Qualitative analysis & small scale separations
Gas-liquid (GLC)	gas	liquid	partition Qualitative & Qualitative Analysis
Paper	liquid	liquid	partition	Qualitative & Qualitative analysis of polar and ionic cmpds

In paper and TLC, you determine the Rf value of a substance, which is the decimal value of the distance that a substance has moved (from its starting point to its final point after a fixed period of time) divided by the decimal value of the distance the solvent has moved. The solvent is measured from the same starting point as that of the substance, to the solvent fronts (the furthest the solvent has moved).

$$R_f = \frac{B-A}{C-A} = \frac{\text{distace spot moved}}{\text{distace spot moved}}$$

WHEN THE STATIONARY PHASE IS A SOLID

It involves the equilibrium between the adsorption of the compound on the solid surface by adhesive forces (intermolecular bonding between the substances and the molecules of the solid phase) and its solution in the liquid phase (also intermolecular bonding but between the substances and the molecules of the liquid phase).

- The liquid passes over the surface of the solid phase in one direction. In column chromatography the liquid normally moves from top to bottom (making use of gravity).
- In thin-layer (TLC) the liquid moves against gravity by moving by capillary action between grains or particles of the solid phase.

WHEN THE STATIONARY PHASE IS A LIQUID

It deals wih equilibrium of partition between the two fluids (the stationary liquid phase and either the liquid or gas of the mobile phase). It involves relatives strength of intermolecular bonding with the substance with each of the fluids in the liquid or gas phase.

- In gas-liquid chromatography (GLC) the stationary phase is a liquid (silicon oil, carbowax, etc.) that is coating an inert solid that is packed in a tube (glass, stainless steel or copper).
- In paper chromatography the liquid stationary phase is generally water that is adsorbed onto teh fibers of the paper. It is another liquid that has different physical properties and constant from those of the staionary phase, even though the liquids themselves may be mutually soluble in one another.
- In GLC the mobile phase is a gas (generally He is used although any inert gas will do).

ADSORPTION CHROMATOGRAPHY

The principle of Adsorption chromatography is based upon th separation of compounds by selective adsorption-desorption on a solid matrix present within a column through which the mixture passes. The separation of proteins is based on the differential affinities of differential proteins for the solid matrix.

Adsorption chromatography is of two types commonly used for the separation of proteins. Separation can be carried out using either an open column or high-pressure liquid chromatography.

AFFINITY AND ION-EXCHANGE CHROMATOGRAPHY

Both ion-exchange and affinity chromatography are commonly used to separate proteins and aminoacids in the laboratory. They are used less commonly for commercial separations because they are not suitable for rapidly separating large volumes and are relatively expensive.

(i) Ion Exchage Chromatography

It is used on the reversible adsorption-desorption of ions in solution to a charged solid matrix or polymer network and is the most commonly used technique for seperation of protein. The charged matrix can be called:

Anion-exchaner: It sis a positively charged matrix and it binds negatively charged ions (anions).

Cation-exhanger: A negatively charged matrix and binds positively charged ions (cations).

To favor maximum binding of the protein to the ion-exchange column the pH and ionic strength are accordingly adjusted. This is the reason why contaminating proteins bind less strongly and therefore pass more rapidly through the column. The protein of interest is then eluted using another buffer solution which favors its desorption from teh column (e.g., different pH or ionic strength).

(ii) Affinity Chromatography

This process of sepeartion of proteins uses a stationary phase that consists of a ligand covalently bond to a solid support. a ligand is a molecules that has a highly specific and unique reversible affinity for a particular protein. The sample to be analyzed is passed through the column and the protein of interest is then eluted using a buffer solution which favors its desoption from the column.

This technique is the most efficient means of separating anindividual protein from a mixture of proteins, but it is the most expensive,because of the need to have columns with specific ligands bound to them.

(iii) Thin Layer Chromatography (TLC)

In TLC, the stationary phase is a silica layer on a glass plate, or just plain (as you will have used at school). The mobile phase can be anything, but is often methanol, water or toluene. The mobile phase moves by capillary action, rather than by pumping, which is the norm for chromatography. The doctor is usually the eye: you can see the spots, althouh you may need to view the plate under UV light. You can use flurescent plates (for nonfluorescent compounds, which will appear as dark spots), or nonfluorescent plates (for fluorescent compounds, which all appear as bright spots). None that TLC is denaturing to some compounds, as the solvents are quite nasty, and the dechnique involves spotting compounds onto a plate, letting them dry, then running them in a tank full of air (i.e. it's rather oxiddising). TLC is useless for proteins therefore

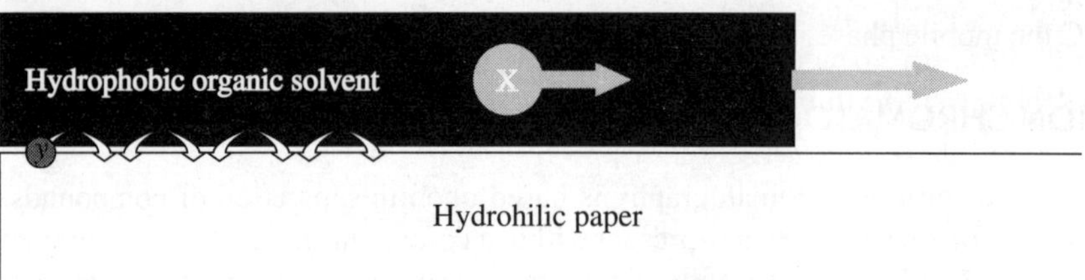

TLC separates molecules based on thier partition between a solvent and a silica layer. The ratio of the distance the compounds moves (i.e. where the middle of its spot is) to the distance the solvent front moves is called the R_f value, and is chracteristics under a given set of conditions for a particular compound. TLC can be used to prepare up large amounts of a compound, rather than for analysing exactly how much is there. This is called *preparative-TLC*

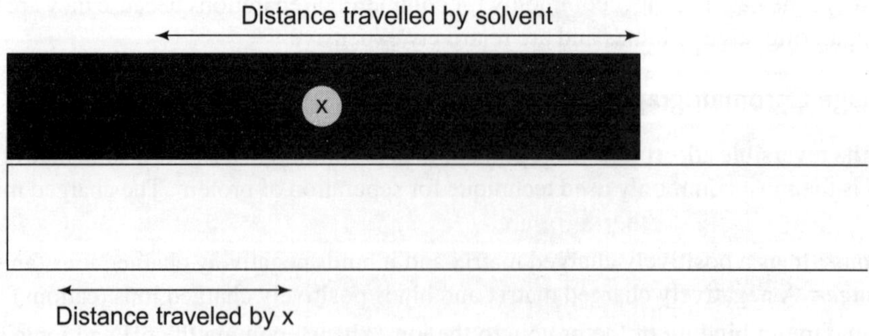

Fig. 53 Mechanism of Chromatography

ADVANTAGES OF TLC

TLC is an important tool for rapid separation of bviomolecules in biological research, because it provides greater resolving power, greater speed of separation (30-60 min), wider choices of adsoren matrixes and greater sensitivity (10 and 100 times more sensitive for amino acids and nucleotides, respectively). Moreover, TLC is advantaveous for easy detection of compounds (permits use of even harsh spray reagents such as 50% sulphuric acid) and easier elution of separated compounds from the TLC plate. Thus, TLC combines better resolution of column chromatography and easier manipulation of paper chromatography.

PREPARATION OF THIN LAYER PLATE

The stationary phase is prepared as slurry with water or buffer at 1:2 adn applied to a glass plate or an inert plastic or aluminum sheet, as a thin uniform layer by means of a spreader such as glass rod or pipette or using a TLC applicator. (0.25 mm thickness for analytical separations and 2.5 mm thickness for preparative separations are prepared).
Calcium sulphate $CaSO_4.1/2\ H_2O$, (Gypsum) (10-15%) is incorporated to the adsorbent as binder, as it facilitates the adhesion of the adsorbent to the plate. After application of the adsorbent, the plates are airdried for 10-15 min and then oven-dried for 10-15 min at 100°C - 110°C. This process is also known as activation of the adsorbent. The plates can be used immediately or stored in a desiccator.

APPLICATION OF SAMPLE

Draw a line lightly with a pencil (do not use microtip pencil as it makes a cut in the layer) about 1.5-2.0 cm from the bottom. (If the thin layer is too soft to draw a pencil line, place a scale at the bottom and spot at a distance of ~ 1.5 cm. Note down the order. The samples are spotted using capillary tubes at ~ 1.5 cm distance between them. For preparative TLC, the sample is applied as a band across a layer rather than as a spot.

DEVELOPMENT OF CHROMATOGRAPHY PLATE

The chromatography tank is filled with the developing solvent to a depth of ~ 1.5 cm and equilibrated for about 5 hrs. The thin layer plate is placed gently in the tank and allowed to stand for about 60 min. Make sure the spots do not touch the solvent directly. Capillary action cuses th solvent to ascend as in paper chromatography and the separation of compounds takes place. As the solvent front raaches about 1-2 cm from the top of the plate, the plate is removed, solvent, solvent front is marked with a pencil immediately and allowed to air-dry placing the plate uside down.

DETECTION OF SEPARATED COMPOUNDS

Several methods are available to detec the seperated components. The plate can sprayed with 25%-50%conc.sulphuric acid in ethanol and incubated at 100%°C for 10-15 min where the substances appear as brown spots. If the compounds are fluorescent, then they can be detected using an UV light. For unsaturated fatty acids, the plate can be developed in a desiccator with a few crystals of iodine.

For amino acids, the plate can be sprayed with 0.1% ninhydrin in acetone and incubated at 100°C for 5-10 min.

For carbohydrates, the plate can be directly sprayed with 50% sulphuric acid or sprayed with 1% napthoresorcinol in 50% TCA or 2% resorcinol in 50% conc. sulphuric acid (unpublished results).

If the compounds are redioactive, the plate can be subjected to autoradiography.

ASSESSMEMEN OF THE AMOUNT OF A PARTICULAR COMPONENT IN A GIVEN SPOT

The amount of a particular compound present in a given spot can be determined by a number of ways. In ways the case of radio labeled compounds, quantification can be achieved by radiochromatogram scanning, or developing the plate with an X-ray film followed by densitometry. Densitometry can also used to locate and estimate compounds, which adsorb in the UV or visible regions.

Scrapping of the spot into a test tube and eluting the compound with a suitalble solvent may carry out off-plate quantification. The amount of the compound present in the solution can then be determined by standard methods after a brief centrifugation.

Table: Adsorbent for TLC:

Silica gel	-amino acids, peptides, fatty acids, steroids, phospholipids, Glycolipids and plasma lipids.
Aluminium oxide	-amino acids, sterorids, vitamins and small organic molecules
Kieselghur	-amino acids, fatty acids, triglycerides and steroids.
Celite	-steroids
Cellulose powder	-amino acids and nucleotides
Calcium phosphate	-proteins and polynucleotides
Hydroxyapatite	-polypeptides and proteins
Polyethyleimine	-nucleotides and oligonucleotides

55

CHROMATOGRAPHY OF AMINO ACIDS

TLC OF AMINO ACIDS
Solvent system

(1) n- Butanol/Acetic acid / Water (8:2:2)
(2) 96% Ethanol/Water (7:3)
(3) Chloroform/Methanol/17% NH_4OH (2:2:1)

Spray reagent

0.1% solution of ninhydrin in acetone, heat for ~10 min in a 100°C oven.

Elution

Five ml of 80% ethanol or methanol.

TLC of Sugars

Prepare silica gel-G plates in distilled water or in 0.2 M sodium borate buffer, pH 8.0.
Standard sugars: 10 mg/ml in water (load 5x1µl)

Solvent System (v/v)

- Ammonia/ ethyl acetate/ n- propanol/ water (1:1:6:3)
- Isopropanol/Pyridine /Water/Acetic acid (8:8:4:1)
- Ethylacetate/ Isopropanol/ Water/ Pyridine (26:14:7:2)
- Benzene/ acetic acid / Mathanol (20:20:60)
- n-Butanol/ Acetic acid / water (2:1:1)
- n-Butanol Acetic acid/ Diethyl ether/ Water (9:6:3:1)
- Chloroform/ Acetic acid/ Water (60:70:10)
- n-Butanol/ ethanol/ Water (5:3:2),useful for separation of starch, maltose and glucose
- n-Propanol/ ethanol/ water (7:1:2), useful for separation of oligosachharides
- n-Butanol/Acetone/ Water (4:5:1),useful for mono, di tri and tetrasachharides

Spray Reagent

- Depending on the availability, any one of the following spray reagents can be used to detect the spots.
- Mix equal volume of NH_4OH and saturated solution of $AgNO_3$, dilute with methanol to give a final concentration of 0.3 M. Spray the chromatogram and place it in a hot air oven for 10-15 min. Reducing sugars appear as brown spots. Or
- Dissolve 0.2% Resorcinol or 5% Naphthoresorcinol in ethanol and mixed equal volume with 50% TCA or H_2SO_4 and then spray. Place in a hot air oven at ~100°C for 15-20 min.
- Or, Mix 1% p-Nitrophenol and 4% O-phosphoric acid and spray.
- Or, spray with 10% H_2SO_4 in ethanol (most of the sugars give black spots)
- Or, Mix 15 ml 85% phosphoric acid, 2 ml aniline and 2 g diphenylamine in 100 ml of acetone, spray and place it in a hot air oven for 30 min at 60° C.

56

SEPARATION OF AMINO ACIDS BY PAPER CHROMATOGRAPHY

Separation of Amino Acids by Paper Chromatography

Materials required

- **Amino acids**:
  Leucine, proline, histidine, alanine, unknowns
- **Chromatogram paper**
- **Chromatographic Chamber**
- **Developing solvent**
  Acetic acid, iso-propanol + ninhydrin (place in chromatography chambers)

Procedure

- Take a piece (4" x 8") of chromatography paper and draw a light pencil line 3 cm from the longer edge.
- Place an X at 3 cm, 6 cm, 9 cm, 12 cm and 15 cm (as shown).
- While handling the paper, hold it on the edges. It is advisable to handle it as little as possible
- Under the first X place the letters Leu (this is where the known amino acid, leucine, will be applied to the paper), under the second X place the letters Ala(for alanine), under the third X place the letters His(for histidine), under the fourth X place the letters Pro(for proline), under the last X place the number or single letter of the unknown you are assigned by your lab instructor.
- All of these marks must be done in graphite (or lead) pencil. Never use ink. (Fig.)

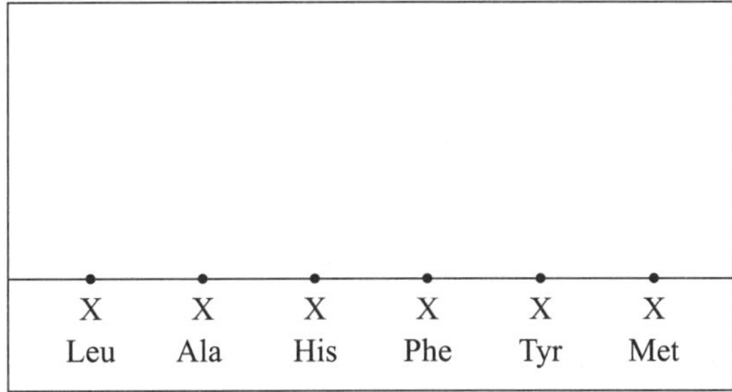

Fig. 54 Preparation of paper for chromatography

2. Put a small spot of the leucine solution on the X that has been labeled Leu. For the remaining X's repeat the same procedure with different amino acids as mentioned with separate capillaries. Never use the same capillary for different solutions; this might contaminate each spot with the previous substance.

3. Allow the spots to dry. Carefully put the paper into the chromatography chambers, ascertaining that the starting line is not below the level of the solvent in the chromatography chamber. The developing solution should be poured beforehand in the chamber. Allow the chromatogram to develop at least one hour. The amino acids will move up the paper by capillary action at different rates depending on their relative solubilities in the developing solvent and the water on the cellulose of the paper.

4. Remove the paper after the solvent has raised a sufficient amount, and mark with a pencil line to the farthest position on the paper reached by the solvent. Immediately take the paper to the drying chambers and dry it gently with a hair dryer.

5. The amino acids react with the ninhydrin (in the developing solution) spots. Remove the chromatography paper from the drying chamber once the spots are developed sufficiently. Circle the spots and measure the distance each has traveled compared to the distance moved by the solvent. Record this on the data sheet. These are the Rf values. Also note the color of the spots.

6. From the Rf values of the known amino acid solutions and the Rf values of the spots of your unknown, one can deduce what amino acids are present.

R_f values:

Ala	-	0.215
His	-	0.319
Pro	-	0.233
Met	-	0.503
Lys	-	0.184
Glu	-	0.288
Tyr	-	0.423

Fig. 55 Paper and Thin Layer Chromatography Apparatus (The principle of separation in paper and thin layer chromatography is partitioning. Particles are separated based on the interactions between two non-miscible liquid phases)

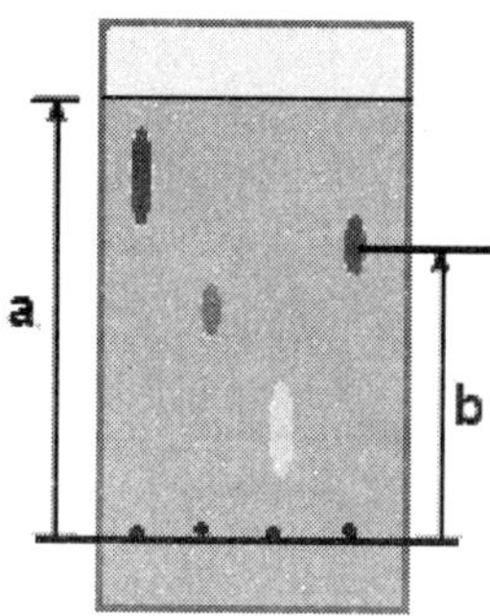

Fig. 56 A typical paper chromatogram

Final Data SheetPart I
Unknown number
Color
Distance solvent rose
Distance leucine rose

Distance alanine rose
Distance histidine rose
Distance proline rose
Amino acids contained in unknown:

Part II
1. Known Sample
Retention time   % Composition
Methanol
Hexane
n-Propanol
2. Unknown sample number: **Components present in mixture**
1.
2.

57

CHROMATOGRAPHIC SEPARATION OF SUGARS

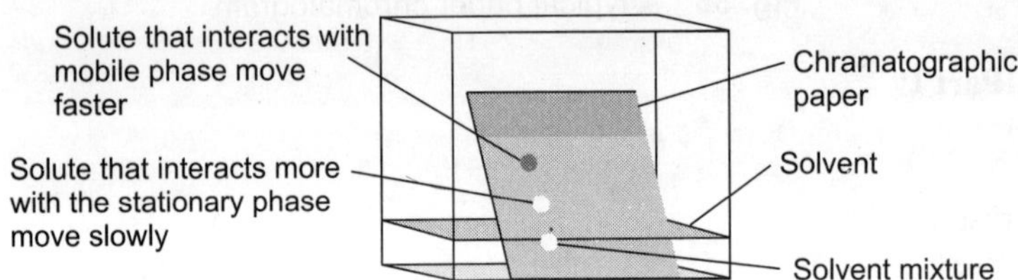

Fig. 57 Chromatographic separation of sugars

In this type of chromatography separation is due to differential partition of solutes between two liquid phases .One liquid phase is bound to the porous medium for example, the water bound in the cellulose paper; this phase is referred to as, the stationary phase. The other liquid phase, the mobile phase flows along the porous medium. As the mobile phase flows over the solute mixture, the individual solutes partition themselves between the aqueous stationary phase and the organic mobile phase relative to their solubilities in the two phases. The more soluble a solute in the mobile phase, the faster it will travel along the paper, and conversely, the mobile phase must be a mixture in which the compounds to be separated are soluble or partially soluble.

In paper chromatography solute or solute mixture is spotted in solution along a base line on a sheet of filter paper (Whatman No. 1).The mobile phase (solvent) is allowed to flow over the spots either ascending the paper by capillary action or descending the paper by gravity. The separation is measured in terms of a unit called Rf (relative rates of flow) with respect to the solvent front. The Rf value can be calculated by the formula mentioned below:

The Rf value of a compound is constant in a particular solvent system under identical conditions, like temperature, pH, etc.

Because most compounds are colorless the spots are visualized after separation by specific reagent. The location reagent is applied by spraying the paper or rapidly dipping it in a solution of the reagent in a volatile solvent. Viewing under ultraviolet light is also useful since some compound which absorbs it strongly show up as dark spots against the florescent background of the paper.

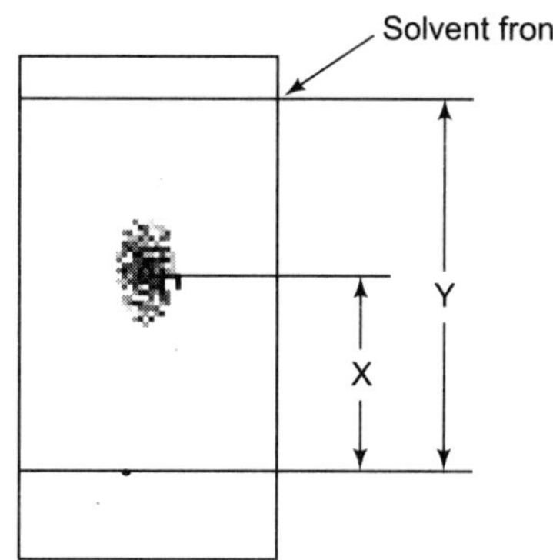

$$Rf = \frac{X}{Y} = \frac{\text{Distance moved by the sample}}{\text{Distance moved by the solvent}}$$

Materials

- **Whatman No:1 filter paper**
- **Solvents**
  a. Water saturated phenol + 1%+ ammonia
  b. n-butanol-acetic acid-water (4:1:5 v/v)
  c. isopropanol-pyridine-water-acetic acid (8:8:4:1 v/v)

- **Spray reagents;**

  A. Ammoniacal silver nitrate:

  Mix equal volumes of NH_4OH to a saturated solution of $AgNO_3$ and dilute to give a final concentration of 0.3M.

  B. Alkaline permanganate:

  Prepare aqueous solution of $KMNO_4$ (1%) containing 2 % Na_2CO_3.

  C. Aniline diphenylamine reagent:

  Mix the reagents in the followinf composition
  - 5 volumes of 1% aniline and
  - 5 volumes of 1% diphenylamine in acetone with 1 volume of 85% phosphoric acid.

D.Resorcinol reagent:

Mix 1% ethanolic solution of resorcinol and 0.2N HCl (1:1 v/v).

Procedure

1. Place sufficient solvent into the bottom of the tank. Cover the led and allow the tank to be saturated with the solvent.
2. Take a sheet of Whatman No.1 chromatography paper (about 9×10 cm) and place it on a piece of clean paper.
3. Draw a fine line with a pencil along the width of the paper and about 1.5 cm from the lower edge.
4. along this line place four equally spaced (about 2cm apart) small circles with a pencil.
5. Label the paper at the top with the name of each of the sugars and label the last unknown.

6. Use a fine capillary or tooth pick to place the drops of the solutions of the sugars, glucose, fructose, maltose, lactose and the mixture.
7. After spotting, dry the paper with hot air dryer for one minute, repeat this step again.
8. Place the spotted paper in the chromatographic tank and make the development by using the ascending technique.
9. Close the tank with lid, allow the solvent to flow for about 30-45 minutes.
10. Remove the paper and immediately mark the position of the solvent front with a pencil.
11. After the chromatogram has dried, spray the paper with the locating reagent.
   - If the developed chromatogram is sprayed with Ammoniacal Silver Nitrate and dry it for 5-10 minutes, and the reducing sugars will appear as brown spots.
   - In case it is sprayed with Alkaline Permanganate and keep the chromatograms in the oven at 100°C for a few minutes, the sugar spots appear as yellow spots in purple background.
   - The dried chromatograms can also be sprayed with Aniline Diphenylamine reagent; the spots are visualized by heating the paper at 100°C for a few minutes.
   - Spray the dried chromatograms with Resorcinol reagent and visualize spots by heating at 90°C.
12. Carefully encircle the position of each spot with pencil.
13. The Rf value for each spot is calculated and also for the spots the mixture contained.

Note: *You need to put the paper on the hot plate at low temperature or expose it to the hot air dryer, until the colored spots appear. The colors are stable for some weeks if kept in the dark and away from acid vapors.*

General summary of the behavior of the various sugars to these reagents are given below:

d	c	b	a	Sugars
pink	+	+	+	Aldohexoses
red	+	+	+	Ketohexoses
Blue, green	+	+	+	Aldopentoses
–	+	+	+	Ketopentoses
–	+	+	–	Deoxy sugars
–	–	–	+	Glycosides
–	+	+	+	Amino sugars

The table below Rf values of some sugars in the solvents previously mentioned. They are only for comparative purposes, since Rf varies with physical parameters.

Solvent c	Solvent b	Solvent a	Sugar
0.64	0.18	0.39	Glucose
0.62	0.16	0.44	Galactose
0.68	0.25	0.51	Fructose
0.76	0.31	0.59	Ribose
-	-	0.73	Deoxy ribose
0.46	0.09	0.38	Lactose
0.50	0.11	0.36	Maltose
0.62	0.14	0.39	Sucrose

Results Sheet

1. Draw a sketch of your chromatogram.
2. Calculate Rf values for each spot of the mixture being separated.
3. By comparing the Rf values of the mixture along with those for the standards, state what sugars does this mixture contain?

HAEMATOLOGY

58

ESTIMATION OF TLC

THEORY

The number of leukocytes (white blood cells) contained in 1 litre of blood is called leukocyte number concentration or leukocyte or white blood cell count, expressed as cells per cubic mm. Such estimation is essential to detect whether a person is suffering from some sort of blood related disorder or any disease, when the number of leukocytes in the blood is altered..

NORMAL RANGE

Case	SI Units (Cells x 10^9 / litre)	Traditional Unite (Cells / mm)
1. Men & women	4 – 10	4000 –10000
2. Children of 10years	4 – 10	4000 –10000
3. Children of 3years	4 – 11	4000 –11000
4. Infants (3-9 months)	4 – 15	4000 –15000
5. Newborn infants	10 – 20	10000 –20000

HIGH VALUES

Increase in the total number of leucocytes, due to pyogenic bacterial infection, leads to *leucocytosis*. In leukemia, leukocyte number concentration can reach up to 50×10^9 / litre to 400×10^9 / litre (i.e. 50,000 /mm^3 to 400,000/mm^3) or even more.

LOW VALUES

A decrease in the total number of leukocyte leads to leucopoenia, this can occur with infections like typhoid and malaria. This condition also occurs after treatment with certain drugs.

REQUIREMENT

- Blood pipette graduated to the 50ìl (0.05ml or 50mm^3) mark with rubber tubing and mouthpiece.
- 1ml graduated pipette.
- Haemocytometer, with counting chamber and special cover glass.
- Diluting fluid.

- Hand Tally counter (optional)
- nticoagulant
- Blood

DILUTING FLUID

It contains a weak acid (to lyse erythrocytes) and a dye (for staining the nuclei of the leukocytes), having the following composition:

- Glacial Acetic Acid ($Na_2\ SO_4$)   = 1.5 – 3ml
- Distilled Water   = 96.5 – 97 ml
- Aqueous Solution of Methylene Blue or gentian Violet   = A few drops

Note: Diluting fluid should be filtered before use.

METHODOLOGY

- Take 0.95ml of diluting fluid in a small bottle with the help of 1ml-graduated pipette.
- Draw venous or capillary blood to the 0.05ml mark of the blood pipette, avoiding taking in any bubble. In case of venous blood mix thoroughly with anticoagulant before pipetting.
- Wipe the mouth of the pipette with absorbent paper or cotton, ensuring that the blood level is still touching the mark.
- Blow the blood into the bottle containing diluting fluid; the dilution of blood is 1 in 20.
- Rinse the pipette by drawing in and blowing out diluting fluid twice or thrice.
- Label the bottle.
- Attach the cover glass to the counting chamber, pressing it carefully into place.
- Using a Pasteur pipette fill the counting chamber, taking care not to overfill the ruled area. In case there is an overflow start the procedure again.
- Leave the Haemocytometer for 2 to 3min. to allow the cells to settle.
- Place the Haemocytometer on the stage of the microscope. Use a x10 objective to count the leukocytes.

TOTAL LEUKOCYTES COUNT

Use of Neubauer Chamber

- Area of chamber   - 9 mm^3
- Depth of chamber   - 0.1mm

Count the cells in an area of 4 mm^2 using the squares numbered 1,3,7, and 9 as shown in the fig. Include in the count the cells seen on the lines of two sides of each square counted, as shown in the fig.

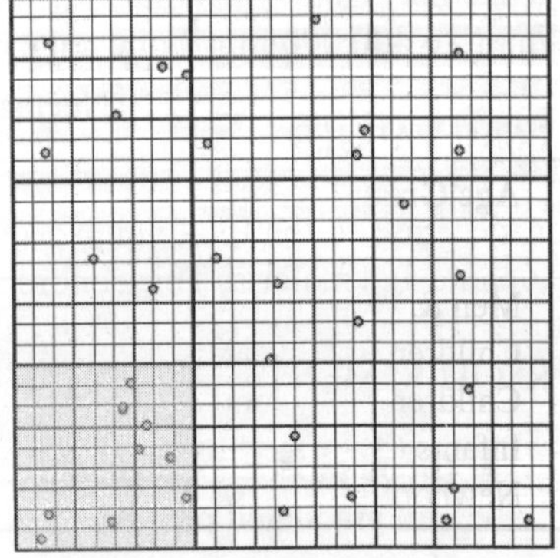

Fig. 58   Neubauer Chamber of Haemocytometer

CALCULATION OF THE NUMBER OF CELLS IN 1 LITRE OF BLOOD

- Multiply the number of cells counted in the four squares by 0.05
- Represent the result as *"number $\times 10^3$ / litre"*

For Example

If the number of cells counted	=	198
Cells in I litre	=	$(198 \times 0.05) \times 10^9$
Result	=	9.9×10^9/litre

EXPLANATION OF THE CALCULATION

Each of the four squares in which cells are counted has an area of $1mm^2$; therefore the total area will be $4mm^2$. The chamber has a depth of $0.1mm$, thus the volume in which the cells are counted is $4 \times 0.1 = 0.4$ mm^3. Therefore a division by 4 and multiplication by 10 will give the number of cells in $1mm^3$ of diluted blood. Since the dilution is 1 in 20, multiplication by 20 will give the total number of cells in $1mm^3$ of undiluted blood. There are 1 million (10^6) cubic mm. in 1 litre, so multiplying by 10^6 will give the number of cells per litre of undiluted blood. The overall calculation is as follows:

$$\text{Cells per litre} = \frac{\text{Cells Counted} \times 10 \times 20}{4} \times 10^6$$

$$= \text{Cells Counted} \times 50 \times 10^6$$

$$= \text{Cells Counted} \times 0.05 \times 10^9$$

For Example

If 190 cells are counted in the four squares. The number of cells / mm^3 of undiluted blood is therefore:

$$\frac{190 \times 10 \times 20}{4} \quad (=190 \times 10 \times 20)$$

and the number /litre is $\dfrac{190 \times 10 \times 20}{4} \times 10^6 = 20$

Normal Leukocyte Count (for human blood)

Age Group	SI units (Cells $\times 10^9$/litre)	Normal Range (Cells /mm^3)
Men & Women	4 – 10	4000 – 10 000
Children of 10 years	4 – 10	4000 – 10 000
Children of 3 years	4 – 11	4000 – 11 000
Infants (3 – 9 months)	4 – 15	4000 – 15 000
Newborn infants	10 – 20	10 000 – 12 000

HIGH AND LOW LEUKOCYTE VALUES

- An increase in the total number of leucocytes is called 'Leucocytosis'. This happens due bacterial infection, blood cancer etc. In leukaemia, the leukocyte number reaches a value of 50×10^9/ litre to 400×10^9/litre (50 000/mm^3 to 4000 000/mm^3) or even higher.
- A decrease in the total number of circulating leukocyte is called 'Leukopenia'. This take place due to typhoid fever, malaria or following treatment with certain drugs.

59

ESTIMATION OF TEC

THEORY

The number of erythrocyte (RBC) contained in 1 litre of blood is called the erythrocyte number concentration. It is expressed as the number of cells /mm^3. The blood is diluted in a red cell diluting fluid. The red cells are counted in a counting chamber under the microscope, and the number of cells in every litre of blood is calculated. Patient suffering from anaemia due to red cell loss or red cell haemolysis will have low red cell count, similarly patients who are dehydrated or have polycythaemia will have high red cell concentration.

MATERIALS REQUIRED

- Pipettes.

   (a) Blood pipette (sometime called Sahli Pipette) graduated to the 0.02ml (20 mm^3 or 20ìl) Mark, with rubber tubing and mouthpiece.
   (b) 5ml graduated pipette.

- Counting Chamber (Neubauer Ruled Chamber).
- Diluting Fluid.
- Hand Tally Counter, if Possible.

DILUTING FLUID

Hayem's diluting fluid contains the following components in distilled water

   (Final volume: 200ml):

- Sodium Sulfate (Na_2SO_4)   = 5.0g
- Sodium Chloride (NaCl)   = 1.0g
- Mercuric Chloride ($HgCl_2$)   = 5.0g

Note

The diluting fluid is an isotonic solution that prevents clotting, haemolysis and bacterial growth

METHODOLOGY

1. Using 5 ml-graduated pipettes, pipette out 4.0 ml diluting into a small bottle.
2. Draw bubble free blood to the 0.02ml mark of the blood pipette. In case of venous blood ensure that it is well mixed with anticoagulant.
3. Wipe the outside of the pipette with absorbent paper, ensuring that the blood is still on the mark.
4. Blow the blood into the bottle of the diluting fluid. Rinse the pipette 2 t 3 times by drawing in and blowing out the diluting fluid. The dilution of the blood is now 1 to 200. Label the bottle.
5. Attach the cover slip of the counting chamber and using the Pasteur Pipette fill the two ruled areas of the chamber, taking care not to overfill beyond the ruled areas.
6. Leave the counting chamber for 2 to 3 min. for the cells to settle down.
7. Place the slide on the stage of a compound microscope and using a x 40 objective count the red cells in the center of the chamber.

COUNTING OF THE RED CELLS

Count the cells in an area of 0.2 mm^2, using the squares marked A, B, C, D, and E (as shown). Include in the count the cells on the lines of two sides of each square counted (as shown).

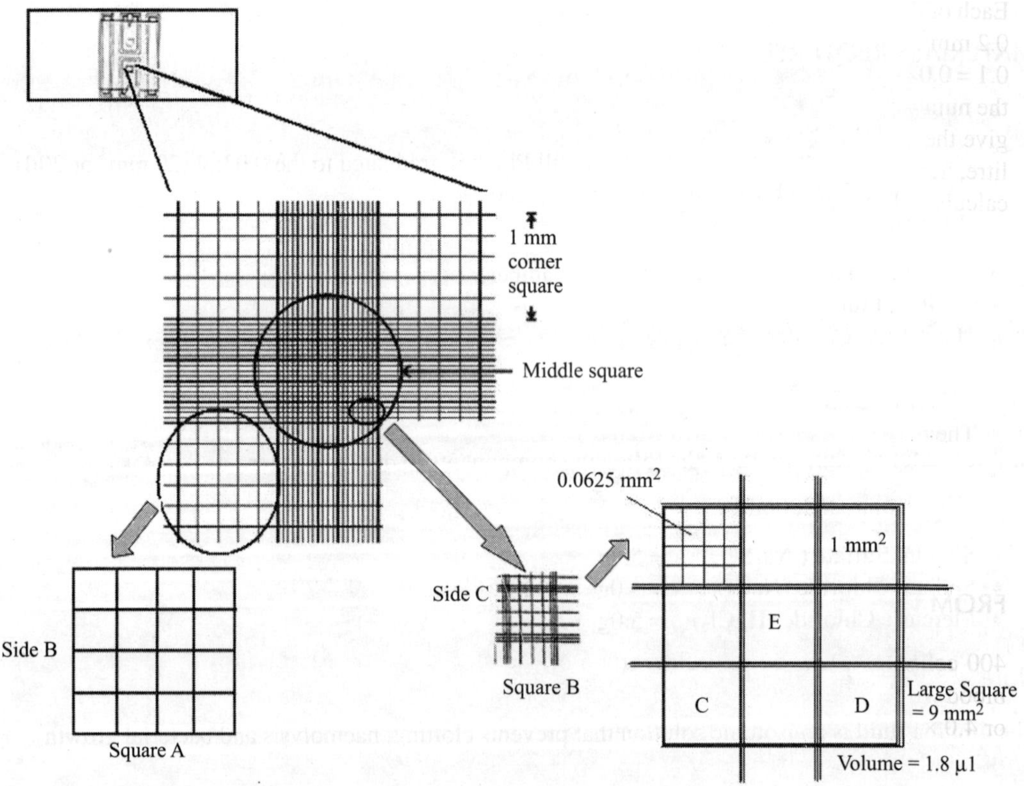

Fig. 59   Counting the RBC in Neubauer ruled Chamber

Calculation of total red cells in 1 litre of blood

- Multiply the number of cells counted in the first group of 5 squares by 0.01
- Repeat the same with the second group of five squares
- Take the average of the two figures
- Represent the result as "number x 10^{12} / litre"

For example

Suppose the number of cells counted in

- First ruled chamber is $\quad = 400$
  Cells per litre $\quad = (400 \times 0.01) \times 10^{12}$
  $\quad = 4.0 \times 10^{12}$
- Second ruled chamber is $\quad = 380$
  Cells per litre $\quad = (380 \times 0.01) \times 10^{12}$
  $\quad = 3.8 \times 10^{12}$
  Average $\quad = 3.9 \times 10^{12}$

EXPLANATION OF CALCULATION

Each of the 5 squares in which the cells were counted has an area of 0.04 mm^2; therefore the total area is 0.2 mm^2. The chamber has a depth of 0.01 mm; hence the volume in which the cells are counted is 0.2 × 0.1 = 0.02 mm^3. Thus dividing the result by 2 and multiplying it by 100 (i.e. multiplying by 50) will give the number of cells in / mm^3 of dilute blood. Because the dilution is 1 in 200, multiplication by 200 will give the number of cells in 1 mm^3 of undiluted blood. Finally, there are 1 million (10^6) cubic mm in 1 litre, hence multiplication by 10^6 will give the number of cells in 1 litre of undiluted blood. The over all calculation is as given below:

$$\text{Cells/mm}^3 = \frac{\text{Cells counted} \times 10000}{1000000}\text{millions}$$

$$= (\text{Cells counted} \times 0.01)\text{ millions}$$

$$= (\text{Cells counted} \times 0.01) \times 10^6$$

The number of cells in 1 litre will be 10^6 times this number, therefore:

$$\text{Cells / litre} = \text{Cells counted} \times 0.01 \times 10^6 \times 10^6$$

$$= \text{Cells counted} \times 0.01 \times 10^{12}$$

FROM THE ABOVE EXAMPLES

400 cells are counted in the first group of five squares. The number of cells in 1mm^3 of undiluted blood will be $400 \times 0.01 \times 10^6 = 3.9$ million. The concentration of undiluted blood is $400 \times 0.01 \times 10^{12}$ or 4.0×10^{12}/ litre.

Experimental Biology

Normal Erythrocyte Count (for human blood)

Age Group	SI units (Cells $\times 10^9$/litre)	Normal Range (Millions of Cells /mm^3)
Men	4.5 – 5.5	4.5 – 5.5
Women	4.0 – 5.0	4.0 – 5.0
Children of 4 years	4.2 – 5.2	4.2 – 5.2
Infants (1 – 6 months)	3.8 – 5.2	3.8 – 5.2
Newborn infants	5.0 – 6.0	5.0 – 6.0

60

HAEMOGLOBIN ESTIMATION

A. CYANMETHAEMOGLOBIN PHOTOMETRIC METHOD

THEORY

Haemoglobin is the blood pigment, composed of iron – porphyrin rings, called *haem,* attached to 4 protein chains, called globin, contained in the erythrocytes. It carries oxygen to the tissue cells of the body. The term haemoglobin was first coined by Hoppe-Seyler in 1863 (Gk. *Haema,* blood; L. *globus,* sphere).

MEASUREMENT OF HAEMOGLOBIN

Haemoglobin concentration is expressed in millimole / litre (mmol /litre) in SI unit. It is necessary to specify the chemical structure to which it applies when this unit is used. In this case the term "haemoglobin (Fe) should be used. Sometimes it is expressed in "gram per litre" (g/L). In this case the simple term "haemoglobin" is sufficient. Values in gram per litre can be converted to millimole per litre by multiplying by 0.062.

For example: Haemoglobin 160 g/l $\times$ 0.062 = Haemoglobin (Fe) 9.92 mmol/l

(Gram / litre is 10 times larger than "gram/100ml", therefore 160g/l = 16.0g/100ml)

PRINCIPLE OF CYANMETHAEMOGLOBIN PHOTOMETRIC METHOD

The blood is diluted in Drabkin diluting fluid, which haemolysis the erythrocytes, converting the haemoglobin into Cyanmethaemoglobin. The solution is analysed in spectrophotometer or colorimeter. Its absorbance is proportional to the amount of haemoglobin present in the blood. Cyanmethaemoglobin photoelectric method gives one of the most accurate haemoglobin estimations.

MATERIALS REQUIRED

- Colorimeter or Spectrophotometer
- Pipettes:
    - (a) Blood pipette (sometime called Sahli Pipette) graduated to the 0.02ml (20 mm^3 or 20ìl) Mark, with rubber tubing and mouthpiece.
    - (b) 5ml graduated pipette.
- Test tubes
- Drabkin diluting fluid
- Cyanmethaemoglobin Standard

Note

1. Drabkin reagent can be bought in tablet or powder forms to be dissolved in 1 litre of distilled water to yield diluting fluid. The solution can be stored at room temperature for one month in a brown bottle. It should not be allowed to freeze as this can result in decolourization with reduction of the ferricyanide.
2. Cyanmethaemoglobin (reference) standard can also be bought. The concentration of the commercially available solution is usually given in milligram/ 100ml (or mg%).

CALIBRATION OF THE COLORIMETER

A calibration curve has to be prepared before the colorimeter can be used for Hb estimations. From such a curve a graph can be prepared and a table made for the Hb values. The steps are as follows:

1. Calculate the Hb value of the reference (standard) solution in grams / litre by using the following formula:

$$\frac{\text{Concentration in mg}/100\text{ml} \times 10}{1000} \times 251$$

Note

10 is the factor for converting 100ml into 1 litre. 1000 is the factor for converting mg to grams, and 251 is the dilution factor when 0.02ml of blood is diluted with 5ml of Drabkin diluting fluid. Because the value of $10 \times 251 / 1000$ comes to nearly 2.5, the above formula can be simplified as follows:

[Hb value of reference (standard) solution in grams / litre = Concentration in mg/100ml $\times$ 2.5]

For example

Concentration of the reference (standard) solution	= 70mg/100ml
Haemoglobin (Hb) value	= 70 x 2.5
	= 175g/litre

   (If the dilution is 1 in 200, i.e. 0.02ml of blood and 4ml of Drabkin dilution fluid is used, the multiply by 2.0 instead of 2.5).

2. Preparation of a serial dilution of the standard solution:

   - Prepare 4 tubes and label them 1 to 4
   - Pipette into each tube as follows:

Test tube number	1	2	3	4
Standard solution	4.0ml	2.0ml	1.3ml	1.0ml
Drabkin fluid	–	2.0ml	2.7ml	3.0ml
Dilution of standard	Undiluted	1 in 2	1 in 3	1 in 4

3. Mix and allow to stand for a few minutes.
4. In the colorimeter read the dilutions as follows:

Standard dilution	Haemoglobin concentration (g/l)	Absorbance
Undiluted	150	35.0
1 in 2	150 / 2 = 75	17.5
1 in 3	150 / 3 = 50	11.5
1 in 4	150 / 4 = 37.5	8.5

- Place the green filter in the colorimeter or set the wavelength at 540nm.
- Fill a colorimeter cuvette with Drabkin dilution fluid and place it in the colorimeter.
- Set zero the colorimeter
- Read the dilution of the test tubes 1 to 4 using the colorimeter cuvette.
- Make sure the reading returns to zero between each reading with Drabkin fluid.

5. Prepare a graph, by plotting the colorimeter readings of the diluted standard solutions against their concentrations.

For example

Standard solution (Reference solution) concentration calculated as 15g/litre

6. Prepare a table of Hb values from the graph, from 20 - 180 g /litre.

HAEMOGLOBIN ESTIMATION OF THE SAMPLE BLOOD

1. Using 5ml-graduated pipette, pipette out 5.0 ml Drabkin diluting into a test tube .
2. Draw bubble free blood to the 0.02ml mark of the blood pipette. In case of venous blood ensure that it is well mixed with anticoagulant.
3. Wipe the outside of the pipette with absorbent paper, ensuring that the blood is still on the mark.
4. Blow the blood into the bottle of the diluting fluid. Rinse the pipette 2 to 3 times by drawing in and blowing out the diluting fluid.
5. Mix the contents of the test tube and leave it for 5 min.
6. Zero the colorimeter, using Drabkin diluting fluid as the blank solution. Read the absorbance of the sample blood using the colorimeter cuvette.
7. Calculate the Hb in g /litre by using the table prepared from the calibration curve.

Age Group	Haemoglobin (fe) mmol / l	Haemoglobin g /litre
Children at birth	8.4 – 12.1	136 – 196
Children at 1 year	7.0 – 8.1	113 – 130
Children, 10 –12 years	7.1 – 9.2	115 – 148
Women	7.1 – 10.2	115 – 165
Men	8.1 – 11.2	130 – 180

Normal Range (for human blood):

B. HAEMOGLOBIN ESTIMATION BY SAHLI METHOD

Theory

When the blood is diluted in an acid solution, the haemoglobin is converted to acid haematin. This test solution is can be matched against a coloured glass reference. This method is not an accurate way of estimating Hb concentration because not all forms of haemoglobin is converted to acid haematin, also the brown colour of the standard is not a true match for an acid haematin solution.

MATERIALS REQUIRED

- Sahli haemoglobinometer
- Sahli pipette (graduated to 20mm^3, i.e. 0.02ml, or 20ìl)
- Small glass rod
- Dropper
- Absorbent paper
- 0.1mol / litre (0.1N) Hydrochloric acid (HCl)

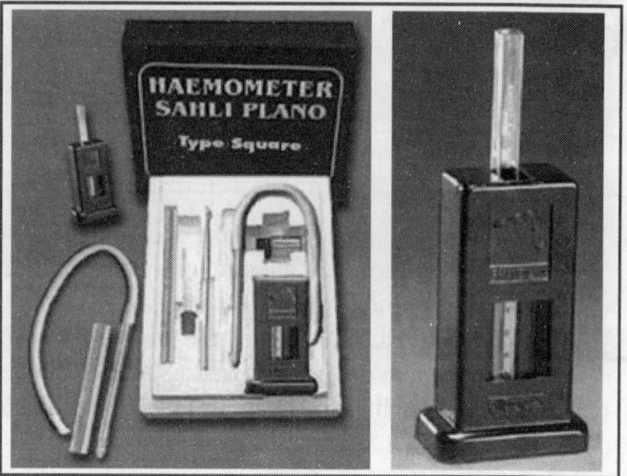

Fig. 60 Haemometer

METHODOLOGY

1. Fill the graduated tube with 0.1 mol /litre HCl to the 20 mark (or the mark 3g /100ml)
2. Draw bubble free sample blood to the 0.02 mark of the Sahli pipette.
3. In case of venous blood ensure that it is well mixed with anticoagulant.
4. Wipe the outside of the pipette with absorbent paper, ensuring that the blood is still on the mark.
5. Blow the blood into the graduated tube of the acid solution. Rinse the pipette 2 to 3 times by drawing in and blowing out the acid solution.
6. Allow it to stand for 5min.
7. Place the graduated tube in the haemoglobinometer and compare the colour of the tube with the colour of the reference tube.
8. Continue to dilute the blood with 0.1mol/litre HCl or distilled water drop by drop and stirring it with the glass rod till the colour matches the reference tube.
9. Note the mark reached. This will give the Hb concentration either in g/100ml or in % of normal.
10. To convert g/100ml to g/litre, multiply by 10. While to convert % to g/litre multiply by 1.46.

For example

- 13.7g/100ml × 10 = 137g/litre
- 87% × 1.46 = 127g/litre

61

ESTIMATION OF HAEMATOCRIT / PACKED CELL VOLUME (PCV) / ERYTHROCYTE VOLUME FRACTION

THEORY

The total volume of the erythrocyte in a blood sample divided by the volume of blood taken is called Estimation of Haematocrit Value or Packed Cell Volume (PCV) or Erythrocyte Volume Fraction. For example, let us suppose the volume of the erythrocyte in one litre of blood is 550 ml, the haematocrit value is 470ml / 1000ml = 0.47 (since the fraction is ml divided by ml, the unit "ml" cancels out, and the result is expressed in simple decimal fraction without any unit. The remaining of the blood is composed of plasma and very little leukocytes. If the leukocyte amount is ignored, then the plasma volume fraction will be 530ml / 1000ml = 0.53 (The haematocrit value + plasma volume fraction = 1). It is thus a measure of the proportion of the RBC to Plasma. It is used in the estimation of mean cell haemoglobin concentration (MCHC). It is also of diagnostic significance for patients suffering from dehydration, shock ,burns, anaemia, etc.

Before the introduction of SI unit the PCV or the Haematocrit value used to be designated by %. Thus in the above example the PCV will be 47% instead of .47.

PRINCIPLE

The sample blood, mixed with anticoagulant, is taken in a graduated tube and then centrifuged to pack the erythrocytes. The level of the column of the erythrocytes is then noted directly from the graduated tube.

MATERIALS REQUIRED

- Electric centrifuge
- Wintrobe centrifuge tubes, with following dimensions:

(a)	Bore	=	0.6cm
(b)	Length	=	9.5cm
(c)	Calibrations	=	0-100

- Pasteur Pipette with rubber teat

PROCEDURE

1. Collect venous blood in a sterilized bottle and add to it a small amount of anticoagulant EDTA)
2. Fill the graduated tube with blood sample up to 100 mark, ensuring that no air bubble is entrapped.
3. Centrifuge for 30mins. At 2500rpm.
4. After centrifugation, the tube will show 3 distinct layers :

   (a) a column of plasma at the top
   (b) a very thin layer of leukocyte in the middle, and
   (c) a column of RBC at the bottom

5. Note the mark at which the layer of RBC meets the layer of leukocytes. The value obtained is a

Percentage (the packed cell volume or haematocrit); divide by 100 to obtain the erythrocyte volume fraction.

RELATIONSHIP BETWEEN ERYTHROCYTE NUMBER CONCENTRATION AND VOLUME FRACTION

The erythrocyte number concentration (cells $\times 10^{12}$ / litre) has some relationship with the RBC volume fraction. If we designate 'C' for the former, the erythrocyte volume fraction will normally in the range of $(C - 0.2) / 10$ to $(C - 0.4) /10$.

For example if the RBC number concentration is 6×10^{12} / litre, the RBC volume fraction will be in the range of $(6 - 0.2) / 10$ to $(6 - 0.4) / 10$; i.e. $= 0.58$ to 0.56.

[In traditional system, the relationship is identical, but the formula used is different: if 'C' is the RBC count, the PCV (haematocrit) as a percentage will be in the range of $(C \times 10) - 2$ to $(C \times 10) - 4$]

RELATIONSHIP BETWEEN ERYTHROCYTE VOLUME FRACTION (EVF) AND HAEMOGLOBIN CONCENTRATION

When Hb concentration is expressed in gm/litre, the EVF is about 0.003 times the Hb concentration.

For example: If a persons Hb concentration is about 125g/litre, then the EVF will be
$$125 \times 0.003 = 0.375$$

Normal Erythrocyte Volume Fraction and PCV/Haematocrit values:

Normal Values	Erythrocyte Volume Fraction	Packed Cell Volume / Haematocrit value (%)
Men	0.40 – 0.50	40 – 50
Women	0.37 – 0.43	37 – 43
Children (5 years)	0.38 – 0.44	38 – 44
Infants(3 months)	0.35 – 0.40	35 – 40
Newborn babies	0.50 – 0.58	50 – 58

- Patients suffering from anaemia have low value: in men it is lower than 0.4 and in women it is lower than 0.37 (PCV 40% and 37% respectively).
- Higher value is present in patients suffering from loss of plasma, severe burns, dehydration, infant diarrhoea and cholera.

62

ESTIMATION OF MEAN ERYTHROCYTE HAEMOGLOBIN CONCENTRATION (MEHC)

THEORY

It is a measure of the average Hb content of the RBC. It is expressed in grams of Hb / litre or in millimole of Hb (Fe) /litre. It is calculated by dividing the Hb concentration of the sample blood by the Erythrocyte Volume Fraction.

EXAMPLE

1. *If the Hb is expressed in grams of Hb /litre:*

$$Hb = 150g/litre; EVF = 0.43$$

$$MEHC = 150 / 0.43 = 349 \text{ g / litre}$$

2. *If the Hb is expressed in millimole of Hb (Fe) / litre:*

$$Hb (Fe) = 9.3mmole / litre; EVF = 0.43$$

$$MEHC = 9.3/ 0.43 = 21.7 \text{ mmol / litre}$$

[Note: To convert g / l to mmol / l, multiply by 0.06206.
In the above example, $349g / l \times 0.06206 = 21.7mmol / litre$]

Result

MEHC normally lies between two limits:
- Lower limit – Hb 322 g / l or Hb (Fe) 20mmol / l
- Upper limit – Hb 371 g / l or Hb (Fe) 23 mmol / l

When the value is between these two limits, the RBCs are said to be "normochromic" (i.e. less coloured than normal), found in patients suffering from hypo chromic anaemia. The RBCs are never hyper chromic (more coloured than normal), but may increase in volume, so that they may contain increased concentration of Hb than normal. In such a case The MEHC may be as high as 380 g/l [Hb (Fe) =23.6mmol/l].

MEAN CELL HAEMOGLOBIN CONCENTRATION:

Previously the MEHC was referred to as 'Mean Cell Haemoglobin Concentration (MCHC)' and was expressed as a percentage. It is calculated by dividing the Hb concentration of the blood grams / 100ml by the Packed Cell Volume (PCV) as a percentage, and multiply by 100.

Example

Suppose the   Hb concentration   = 15.0g/100ml
                PCV   = 43%
                MCHC   = (15.0 × / 43) × 100 = 34%

[Note: In this system the normal range is 32–36% and the value never exceeds 38%]

63

ESTIMATION OF ERYTHROCYTE SEDIMENTATION RATE (WESTERGREN METHOD)

THEORY

The rate at which the erythrocytes of the blood sediments on there own weight when held in a vertical graduated column is called Erythrocyte Sedimentation Rate (ESR). Any change in the ESR value from the normal value indicates the presence of some organic disease. It is expressed as the sedimentation of RBCs in mm at the end of 1st. hour. The entire process of sedimentation takes place in three stages:

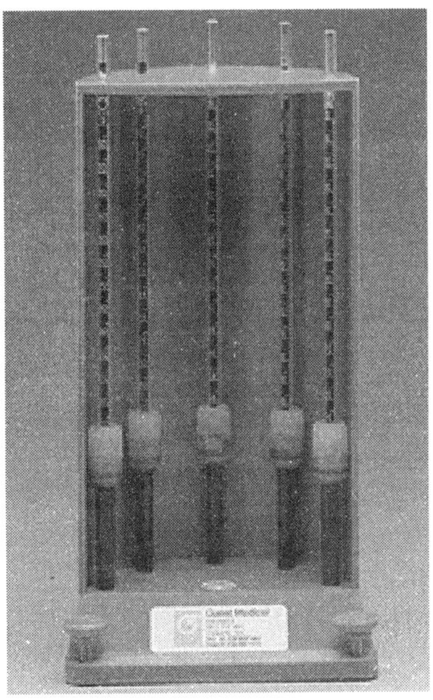

Fig. 61 A Westergren Apparatus

Stage: 1

A brief stage of RBC aggregation (about 10 minutes.) when the rate of sedimentation is slow and Rouleaux formation occurs.

Stage: 2

The rate of sedimentation is constant but at a very high velocity. It lasts for 40 minutes.

Stage: 3

Completion of packing takes place at this stage and the rate is slow. It lasts for 10 minutes. So that at the end of this stage a clear upper layer of plasma is seen floating at the top of the packed RBC column.

MATERIAL REQUIRED

- Westergren ESR tube with following dimensions:

    (a) Internal diameter 2.5mm
    (b) Graduated from 0 to 200mm (sometime marked 1 to 20, 1 corresponding to 10, 2 to 20, etc.)

- Westergren Stand
- Anticoagulant: 38g/l (or 3.8%) trisodium citrate solution or EDTA. Keep in a refrigerator
- 5ml graduated syringe
- Stop watch

METHODOLOGY

1. Take 0.4ml of trisodium citrate solution or EDTA in a tube or bottle
2. Collect 2ml of venous blood into a syringe
3. Add 1.6ml of blood to the bottle containing anticoagulant, so that the final volume is 2.0ml. Shake the bottle gently to mix the content. Measurement of the ESR should begin within 2 hrs of collection of blood.
4. Draw the anticoagulant blood into the Westergren tube up to the 0 mark, without any air bubble.
5. Place the graduated tube in the Westergren Stand. Note that the tube is completely upright and the stand is level.
6. Wait for an hour and then note the height of the column of plasma in mm graduations starting from 0 mark at the top of the tube.

Result

The result is expressed as follows:

ESR ....mm/h

Normal range

Men:	1 – 10mm/h
Women:	3 –14mm/h

Increase in ESR

Increase in ESR value is seen in patients suffering from diseases that produce change in plasma protein or in chronic diseases. Very high ESR is seen in patients suffering from:

- Tuberculosis
- Trypanosomiasis
- Malignant diseases
- Rise in oxygen
- Increased amount of cholesterol
- Increased amount of fibrinogen
- Increased amount Ü -globulin

In normal pregnancy or in high temperature the ESR value is also very high.

DECREASE IN ESR

- Rise in CO_2
- Increase in albumin percentage
- Rise in nucleoprotein value
- Increase in % of lecithin

64

PREPARATION AND STAINING OF THIN BLOOD FILM

PRINCIPLE

A thin blood film is prepared by spreading a small drop of blood uniformly so that there is only one layer of cells. After staining the blood film is analysed for determining or identifying:

- Leukocyte type number fraction
- Abnormal red cells
- Certain parasites

MATERIAL REQUIRED

- Clean grease–free glass slides
- Glass rods (2)
- Measuring cylinder (50ml or 100ml)
- Rack for drying the slides
- Methyl Alcohol
- Leishman Stain
- Distilled water
- $Na_2HPO_4.2H_2O$
- Anhydrous KH_2PO_4
- Stop watch

COLLECTION OF BLOOD SAMPLE

Take the blood from 3rd or 4th finger, from the side. Never collect blood from index finger or thumb, infected fingers or ear, which contain too much monocyte.

PREPARATION OF THE BLOOD FILM

1. Take a drop of blood by touching it with one end of the slide.
2. Place the slide on a flat table and hold with one hand. Using the other hand place the edge of the spreader slide in front of the drop of blood.
3. Draw the spreader slide backwards till it touches the drop of blood.

4. Let the blood spread along the edge of the spreader slide.
5. Push the spreader slide at an angle of 45° to the end of the slide containing the blood. The entire blood should be used up before it reaches the end of the slide. Blood from anaemic patients should be spread more rapidly.
6. The film is satisfactory only when:

   - No lines extending across or down through the film
   - The film is smooth and ragged
   - The film is not very long
   - The film is not very thick
   - There is no holes in the film

7. Dry the film adequately by waving it rapidly in front of a spirit lamp.
8. Mark the dried film with sample number, date, etc.
9. Fix the film with methanol and wrap it in a white sheet of paper.

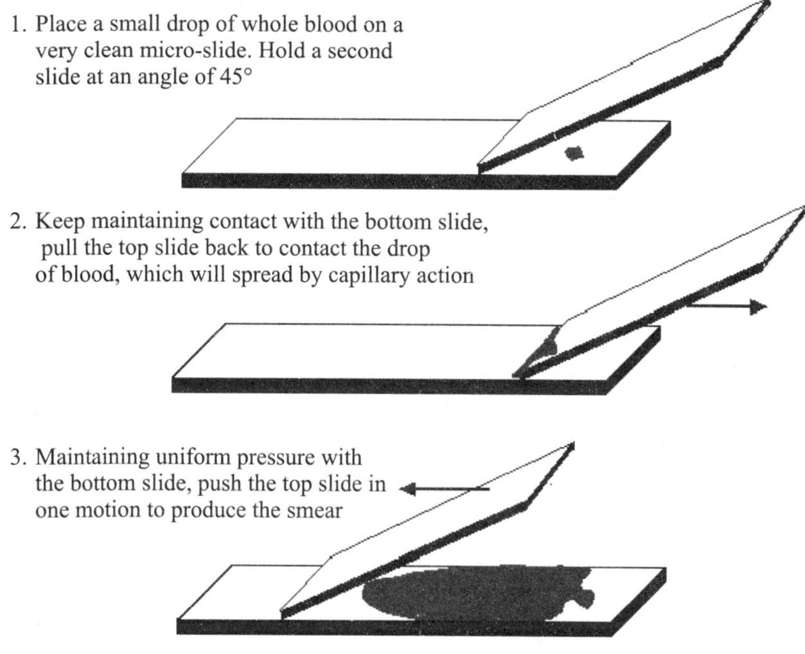

Fig. 62   Preparing a blood film

STAINING THE FILM WITH LEISHMAN STAIN

1. Fix the slide containing the blood film with methanol for 2 – 3 minutes.
2. Prepare a 1 in 3 dilution of Leishman stain using one part of stain and two parts of buffered distilled water. [Note: The buffered water is prepared by mixing 3.76g of $Na_2HPO_4.2H_2O$ and 2.10g of anhydrous KH_2PO_4 in 1000 ml of distilled water, adjust pH between 6.8 – 7.2]. Take 10ml of stain and 20ml of buffered water. It should be used within 24 hrs.

3. Cover the slide with dilute stain for 7 – 10 minutes or more if required.
4. Wash the slide containing the stain thoroughly with buffered water, ensuring that there is no stain deposit in the film.
5. Leave the film for 2 – 3 minutes covered with clean water for the stain to differentiate. The time taken for differentiation is dependent upon the pH of the water used and stains quality, which is important for differentiating the leukocytes with Leishman stain.
6. Drain out the water from the slide and let it dry in the stand.

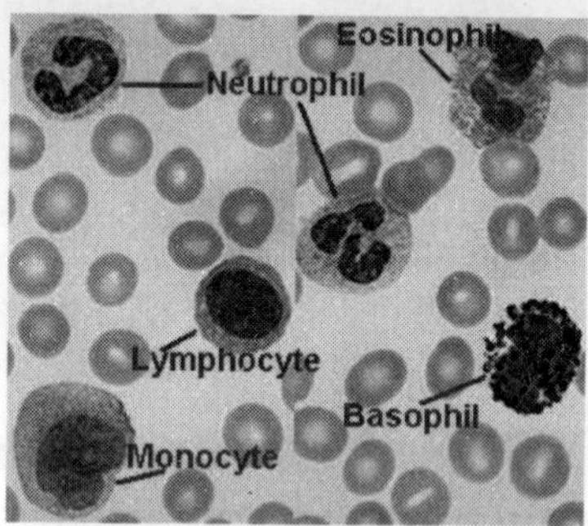

Fig. 63 Human Blood Smear

Result

If the film is stained properly, the following features should be distinct:

Cells	Staining property
Neutrophils	The cytoplasm take up light pink colour and shows small mauve granules.
Eosinophils	The cytoplasm take up light pink colour and shows large red granules.
Monocytes	The cytoplasm stain gray-blue.
Lymphocytes (Large)	The cytoplasm takes up clear blue colour.
Lymphocytes (Small)	The cytoplasm takes up a dark blue colour.
Basophils	The cytoplasm is filled up with many mauve-blue granules.
Erythrocytes	Cells stains pink-red
Platelets	The cells stains mauve-pink

Note: If the slide takes up too much blue colour then in that case rinse twice with alcoholic boric solution (prepare 1 %boric acid in 95 % ethanol), wash immediately with distilled water.

65

EXAMINATION OF LEUKOCYTE AND LEUKOCYTE NUMBER FRACTION

THEORY

The leucocytes are five in number and have different morphology and staining properties. The differences lie in their size, shape of the nucleus, colour of their granules in the cytoplasm, etc. The differential count of leukocytes has diagnostic importance. The proportion of each type is known as leukocyte type number.

PRINCIPLE

Let us suppose 100 leucocytes are counted and the number of different types is noted. The proportion of each type is represented as a decimal fraction. The total of the entire fraction should be 1. For example:

Leukocyte type	Leukocyte Number Fraction	Differential Leukocyte Count
Neutrophil	0.56	56 %
Eosinophils	0.25	25 %
Monocytes	0.12	12 %
Lymphocytes	0.06	6 %
Basophils	0.01	1 %

In the traditional system, the leukocytes number fraction is called "Differential Leucocyte Count" and is represented as %.

MATERIALS REQUIRED

- Compound Microscope with x 10 eyepiece and a x 40 and x 100 oil immersion objective.
- Cover slip.
- Immersion Oil (Cedar wood oil).
- A well spread, stained thin blood film.
- A counting keyboard, if available.

METHODOLOGY

1. Examination of the film

Taking an x 100-oil immersion objective count the leukocytes from the end of the smear. Count the leukocytes by moving from one field to another systematically. Keep a record of the types seen in each field. Count a total of 100 leukocytes. Do not count them in areas where the film is too thick.

2. Examination of the leukocytes

- Compare the shape and size of the leukocytes with that of RBCs.
- Note the shape and size of the nucleus in relation to the total area of the cell – round (R), Lobed (L) or Indented (I).
- Observe appearance of the cytoplasm of the cells:

  Colour of the cytoplasm

  - Colourless
  - Pink
  - Pale blue
  - Dark blue

  Granules in the cytoplasm

  - Neutrophil granules (N) – Small and mauve in colour
  - Eosinophil granules (E) – Large and orange red in colour
  - Azurophil granules (A) – Large and bright reddish purple
  - Basophil granules (B) – very large and mauve-blue granules

- Observe the appearance of the chromatin of the nucleus – densely or faintly stained. Also note the position of the nucleus in the cell – centrally or eccentrically oriented.
- Note the vacuoles (V) in the cytoplasm and nucleoli (N) in the nucleus.

CELL TYPE CHARACTERISTICS

A. Normal Cells

This category includes Polymorphonuclear Neutrophils (P), Lymphocytes (L) and Monocytes (M). While Neutrophils have nucleus with several lobes and granules in the cytoplasm, The lymphocytes and monocytes have a compact nucleus and with or without granules in the cytoplasm.

1. Polymorphonuclear Neutrophils

Size:	12 - 15μm
Shape:	round, well defined
Cytoplasm:	abundant, pinkish
Granules:	small, many, separate and mauve in colour
Nucleus:	several lobes (2 – 5), joined by strands of chromatin, which are deep purple (number of lobes depend upon the age of the cell)

Immature Polymorphonuclear Neutrophils;
Identical to the above except that the nucleus does not have any lobe and is often "S" shaped.

2. Polymorphonuclear Eosinophils

Size:	12 - 15µm
Granules:	large, round, many, closely packed orange red in colour
Nucleus:	usually with two lobes.

3. Polymorphonuclear Basophils (rarest type of granulocyte)

Size:	11 - 13µm
Shape:	round
Granules:	very large, round, many, less closely packed than Eosinophils, dark purple
Nucleus:	usually not seen as it remains covered with granules
Vacuoles:	sometimes small colourless vacuoles in the cytoplasm

4. Small lymphocytes

Size:	7 - 10µm
Shape:	round
Cytoplasm:	Scanty, with no granules, blue
Nucleus:	large, dense, covering the entire cytoplasm, dark purple

5. Large lymphocytes

Size:	10- 15µm
Shape:	round or irregular
Cytoplasm:	abundant, with no granules, light blue
Nucleus:	round, lying on one side of the cell
Granules:	very few, large, azurophilic (dark red)

6. Monocytes

Size:	15 - 25µm, largest of the leukocytes
Shape:	irregular
Granules:	fine, reddish
Nucleus:	variable, kidney shaped, chromatin arranged in strands, mauve
Vacuoles:	usually present in the cytoplasm

(Note: Malaria patients contain brownish black masses, which are Malaria pigment)

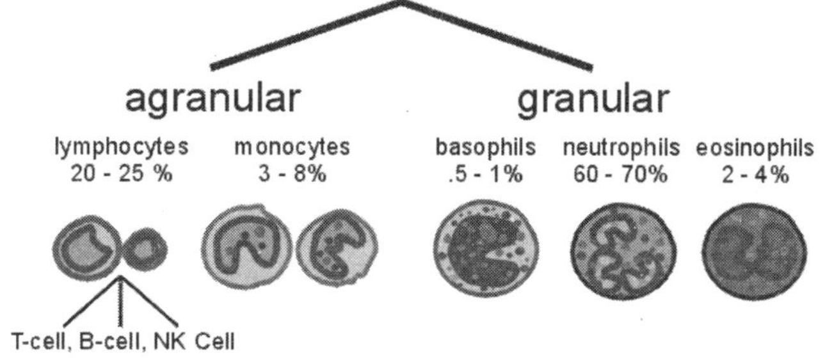

Fig. 64   White blood cell ~ WBC

B. Abnormal or rare cells

Include plasma cells, nucleated red cells or normoblasts, immature granulocyte, hyper segmented Polymorphonuclear Neutrophils, atypical lymphocytes, blast cells, megakaryoctes, etc.

1. Plasma cells

These cells produce antibodies and are frequently seen in the blood of patients suffering from viral and bacterial infections.

Size:	12- 15µm
Shape:	round or oval
Cytoplasm:	dark blue with light staining area around the nucleus.
Nucleus:	round, chromatin clumped, eccentric, wheel-like
Vacuoles:	many, small, and not clearly visible

2. Nucleated red cells or normoblasts

These are often seen in the blood film of patients suffering from anaemia.

Size:	8- 10µm
Shape:	round or irregular
Cytoplasm:	pink or grayish – blue without granules.
Nucleus:	round, chromatin clumped, dense, eccentric, dark staining.

3. Immature Granulocytes

These are often seen in the blood film of patients suffering from severe bacterial infection.

Size:	12- 18µm
Cytoplasm:	pink or pale blue
Nucleus:	round, dark staining.
Granules:	many, large, mauve or dark red

4. Hyper segmented polymorphic neutrophils

Identical to normal neutrophils but the nucleus has larger 5 – 10 lobes. Such cells are present in blood film of patient anaemia due to Vitamin B-112 or folic acid deficiency.

5. Atypical lymphocytes

These can be seen in the blood film of patient suffering from viral infection, tuberculosis, whooping cough, malaria and measles. They have the following characteristics:

Size:	variable, 12- 18µm
Shape:	irregular
Cytoplasm:	dark blue with dark edges.
Nucleus:	round or irregular, eccentric, nucleoli seen

6. Blast Cells

It is the earliest of all type of leukocytes, seen in blood of patients with leukaemia

Size: large, 15-25μm

Cytoplasm: dark blue with clear area around the nucleus.

Nucleus: large, round, pale mauve, with 1 – 5 nucleoli.

METHOD OF COUNTING

Leukocytes can be counted by a number of ways, like using a special counting machine with keys, use of home made bead counter box, or use of a paper and pencil.

1. Counting machine

The kit has a keyboard with keys for each type of leukocyte and the number of each type is noted automatically.

2. Counter Box

It can be locally made containing 5 boxes and labeled as

N =   Neutrophil
E =   Eosinophil
B =   Basophil
L =   Lymphocyte
M =   Monocytes

The 6th. Box containing beads (or bean seeds or wheat grains) for counting

3. By using pencil and paper

Draw a table as shown in the figure and record the different types of leukocytes. When 10 strokes have been made in the first line proceed to the next. Thus, when the 10th line has been completed, you will be able to know that you have counted 100 cells. Add the total for each vertical column.

	N	E	B	L	M
1	1111	11		111	1
2	1111 1	1		111	
3	1111 11			111	
4	1111	1		111	
5	1111 111		1	11	1
6	1111 11	1		11	
7	1111			111	11
8	1111 11	1		11	
9	1111 1	1		111	
10	1111	1		1111	
Total	59	8	1	28	4
Fraction	0.59	0.08	0.01	0.28	0.04

Cell Types	Normal range of leukocytes in different age groups				
	New born	After 4 days	1 – 4 years	10 years	Adults
Polymorphonuclear Neutrophil	0.55–0.65	0.40–0.48	0.36–0.48	0.45–0.55	0.55–0.65
Polymorphonuclear Eosinophil	0.02–0.04	0.02–0.05	0.02–0.05	0.02–0.05	0.02–0.04
Polymorphonuclear Basophil	0.00–0.01	0.00–0.01	0.00–0.01	0.00–0.01	0.00–0.01
Lymphocytes	0.30–0.35	0.40–0.48	0.44–0.54	0.38–0.45	0.25–0.35
Monocyte	0.03–0.06	0.05–0.10	0.03–0.06	0.03–0.06	0.03–0.06

Each different type of leukocyte may be recorded in the form of number concentration (i.e. number of cells /litre) instead of number fraction. The number concentration can be calculated by multiplying the number fraction of a leukocyte by the total leukocyte number concentration.

For example: Leukocyte number concentration $= 6 \times 10^9$/litre
Neutrophil number fraction $= 0.45$
Neutrophil number concentration $= 0.45 \times 6 \times 10^9 = 2.7 \times 10^9$/litre

Note

1. In SI unit the value are given as number fractions. To get the percentage the value should be multiplied by 100.
2. In traditional unit, multiplying the percentage of Neutrophil by the total leukocyte count and dividing by 100 make the calculation. For example:

Total leukocyte $= 5000$/mm^3
% of Neutrophils $= 42\%$
Absolute Neutrophil count $= (42 \times 5000) /100 = 2100$ /mm^3

Abnormal leukocyte count

Disease	Number	Cause
Neutrophilia	Above 0.65	During acute infection
Eosinophilia	Above 0.05	Due to parasitic worm infection and allergy
Lymphocytosis	Above 0.35	Due to virus infection (measles, etc.), chronic infections (malaria, TB, etc.) and some toxic condition
Monocytosis	Above 0.06	Due to bacterial and parasitic infection like malaria, typhoid, kala-azar, etc.
Neutropenia	Below 0.65	Decrease in number may also be caused by certain infections and other diseases

PHYSIOLOGY

66

EFFECT OF TEMPERATURE ON RATE OF HEART BEAT

AIM OF THE EXPERIMENT

To show the effect of temperature on the heart beat rate in toad. Temperature, being an important physical factor, influences the rate of biochemical and physiological processes. The rate of many biochemical processes double with every 10°C rise in temperature (Q_{10}), within a limited range of temperature.

MATERIALS REQUIRED

- 0.6% NaCl solution (isotonic Ringer's solution)
- Chloroform
- Glass jar
- Petri dish
- Beaker
- Thermometer
- Spirit lamp
- Cotton thread and needle
- Stop watch
- Syringe
- 18 gauge needle

METHODOLOGY

- Anesthetize the toad mildly with chloroform.
- Dissect open the thorax and carefully remove the pericardium to observe the heart beat.
- Count and note the heart beat rate under following conditions:

    (a) under normal condition
    (b) bathing with cold and warm Ringer at different temperature – 5°,15°, 25° and 35°C
    (c) 10°C below the room temperature and 10°C above the room temperature

OBSERVATION

Observe and count the heart beat rate under normal and experimental conditions and represent your result in the following tabular form:

Experimental Conditions	Rate of Heart Beat per minute	
	In vitro	*In vivo*
Normal		
5°C		
15°C		
25°C		
35°C		
5°C + Ringer		
15°C + Ringer		
25°C + Ringer		
35°C + Ringer		

Result

(a) The myogenic heart of toad continues to beat even in a denervated (isolated) condition, provided suitable environment is provided, like maintenance of turgid condition by providing isotonic Ringer, suitable pH and temperature

(b) The rate of heartbeat *in vivo* (with both nerves intact) can be counted under normal condition.

(c) The rate of heartbeat of an isolated heart (i.e. denervated and tied to prevent blood loss and keep it filled) can be counted.

(d) Q_{10} value can be calculated by finding out the ratio of the frequency of heart beat by the following formula:

$$Q_{10} = \frac{K_{t-10}}{K_t}$$

Therefore, $$Q_{10} = \left[\frac{K_2}{K_1}\right]^{10/2^{-t}} K_t$$

Where, K_t is the rate at t_1 and K_t the rate at t_2

(e) The response of heart at different temperatures can be observed.

(f) The frequency of *in vitro* beat will differ from *in vivo* beat as the heart muscles mostly depend on the oxygen and glucose and the supply of both are immediately cut off as the heart becomes isolated from the body.

(g) The above experiment enables us to measure the effects of drugs, chemicals, toxins, and neurotransmitters, without the help of any sophisticated instruments.

67

STUDY OF ESTROUS CYCLE
FROM VAGINAL SMEAR OF RAT

THEORY

Sexual cycles in seasonal breeders are highly variable, with a non- breeding or anestrous period during which the animal remains sexually dormant, and a sexually active period or estrous during which the animal becomes sexually receptive both physiologically and psychologically and hence copulation is possible during this period.

During the estrous phase there are drastic changes in the vagina, uterus, ovary, and in the hormonal level of the animal. The histology of the vaginal epithelium changes during different phases of the sexual cycle of the rat, brought about by interaction of various hormones, which too undergo cyclic changes. The epithelium of the vagina is torn down and rebuilt cyclically and fluctuates between stratified epithelium and low cuboidal types. These types can be observed in stained vaginal smears of the rat. The female rats come to estrous every 4 -5 days and are receptive to the males for about 12 hours (range from 9 – 20 hours) and ovulate spontaneously.

The experiment is useful for studying the reproductive and breeding behaviour of the test animals.

MATERIAL REQUIRED

- Female albino rat or Swiss mice as experimental animal of 12 –14 weeks
- Normal saline (0.9% NaCl solution)
- Pasteur pipette with rubber teat
- Microscope with x 10 and x 40 objectives)
- Glass slides
- Cover slips
- Methanol
- Graded ethyl alcohol (50%, 70%, 90% &100%)
- Haematoxyline
- Alcoholic eosine
- DPX mountant
- Glass marking pencil

METHODOLOGY

1. Keep the female rat with adequate amount of food and water for 7 days to get acclimatized
2. Mark the slide putting date, serial number and time.
3. With the help of Pasteur pipette introduce a 0.3 –0.5 ml of normal saline through the vagina of the rat, taking care not to cause any injury
4. Take out the vaginal swab through the same pipette
5. Prepare a thin smear of the swab and allow it to dry
6. Fix the swab by adding a little amount of methanol for 10 minutes
7. Dehydrate and stain the slide by double staining method using haematoxyline – eosine
8. Mount the stained slide with DPX
9. Investigate the smear under microscope using x 10 and x 40 objectives
10. Repeat the experiment at different stages of estrous cycle at regular interval

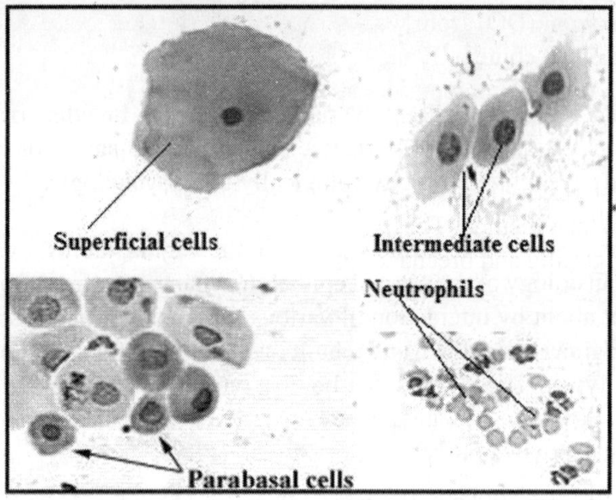

Fig. 65   A Vaginal Smear showing different cells

Result

Following is an account of changes in the ovarian and vaginal histology during the estrous cycle in rat (after Nalbandov, 1958).

Stages	Duration	Changes in Ovary	Type of cells in Vaginal Smear
Diestrus	Half of the whole cycle (50 -55 hrs)	Corpora lutea	Nucleated epithelial cells and leukocytes
Proestrus	About 12 hrs.	Follicles grow faster	Nucleated epithelial
Early estrus	About 12 hrs.	Largest follicles	Cornified squamous cells
Late estrus	About 18 hrs.	Ovulation	Cornified squamous cells
Metaestrus	About 06 hrs.	Corpora lutea formed	Leukocytes among Cornified squamous
Anestrus (or beginning of the diestrus)	–	Functional corpora lutea during early part	Cornified cells disappearing

68

ESTIMATION OF RESPIRATORY RATE IN FISHES

THEORY

Fishes utilize dissolved oxygen (DO) from water and hence the status of an aquatic ecosystem can be evaluated by estimating the DO content. It is an indicator of the water quality. In an aquatic ecosystem the hydrophytes increases the DO content while the fishes and other aquatic organisms decreases it., proportional to the amount of input and output respectively. Thus considering a unit of fish biomass and DO in an air tight jar, it is possible to estimate respiratory metabolic activity of fish relevant to time and space. Quantification of fish respiration is the aim of this experiment which can be conducted by estimating the Do content by Winkler's method. The experiment is to ascertain the optimum environment for the transport of fish seeds, fingerling and breeders for long distance without any problem for DO.

WINKLER'S METHOD

DO reacts with $Mn(OH)_2$ and gets converted to $Mn(OH)_3$ which when acidified liberate equivalent amount of Iodine (I_2) proportional to the DO. The liberated iodine can be estimated by titration against Sodium Thiosulphate ($Na_2S_2O_3$) to calculate the DO.

Respiration by fishes utilizes DO, which eventually decreases in the airtight jar than in the control jar. The difference between the DO of the two will give the amount of DO used by fishes. Conversion to unit dry weight peer unit time and space reveal the rate of respiratory utilization of O_2 by fish.

REACTIONS

$$
\begin{aligned}
1.\ & MnCl_2 + 2KOH && = && Mn(OH)_2 + 2KCl \\
2.\ & Mn(OH)_2 + H_2O + 1/2\ O_2 && = && Mn(OH)_3 \\
3.\ & Mn(OH)_3 + 2KI + 6HCl && = && 2\ MnCl_2 + 2KCL + 2H_2O + I_2
\end{aligned}
$$

MATERIALS REQUIRED

- Manganous chloride (40%)
- Alkaline Iodide
- Starch indicator
- Sodium Thiosulphate
- Chloroform
- Conical flask

- Burette
- Pipettes
- Volumetric flasks
- Beaker
- Sample water
- Test animal – fish (10 – 50g live weight)
- Iron burette stand
- Wash bottle
- Glass jar with lid or air tight plastic container (2 – 3l capacity)
- Rubber tube
- Plastic bucket
- Aquarium

PREPARATION OF SOLUTIONS

(a) Alkaline Iodide (KI)

- Dissolve 500g of NaOH in 500ml of DW
- Dissolve 150g of KI in 500ml of DW
- Mix the two solution together to get the reagent

(b) Starch Indicator

Mix 1g % of starch in warm water, cool and filter it and store it with a few drops of chloroform

(c) Sodium Thiosulphate solution ($Na_2S_2O_3$)

Dissolve 24.82g of $Na_2S_2O_3$, $5H_2O$ in boiled and cooled DW, dilute it to 1litre.

PROCEDURE

The entire experiment is subdivided into five protocols:

1. Setting of control and experimental jars
2. Fixation of water samples
3. Titration
4. Moisture content of fish, and
5. Calculation

CALCULATION

$$
\begin{array}{lll}
Na_2S_2O_3 & : & 1N\ I_2 \ : \ 1N\ O_2 \\
1N\ Na_2S_2O_3 & : & 8g\ O_2 \\
Or\ 1000ml\ of\ 1N\ Na_2S_2O_3 & : & 8mg\ O_2 \\
Or\ 1ml\ of\ 0.01N\ 1N\ Na_2S_2O_3 & : & 0.08mg\ O_2
\end{array}
$$

DO should be calculated per litre depending upon the quantity of $Na_2S_2O_3$ consumed and it is to be estimated for the total volume of the jar or the container. Difference in the DO of the control and the experimental jars indicate the total DO consumed by the fish. Dry weight of the fish is to be determined by oven drying at 85°C. The rate of oxygen consumed is to be calculated in terms of mg DO g^{-1} dry wt, and hr^{-1}

Precautions

1. Take care to prepare the solutions
2. Avoid bubbling while siphoning water from bucket into the jar, this will alter the DO content
3. Handle the fishes carefully, without causing unnecessary stress
4. Titrate carefully and in replicate of 5 – 10 so as to get statistically viable readings
5. Note the time and temperature carefully

SCOPE OF THE EXPERIMENT

- The experiment is designed to determine the metabolic demand of any aquatic animal or plant with altered physical conditions such as temperature, pH, salinity, stress conditions etc.
- DO contents of different water samples can also be compared

APPENDICES

APPENDICES

I. PREFIXES USED IN THE INTERNATIONAL SYSTEM OF UNITS

Prefix	Multiple	Abbreviation	Greek alphabets					
Exa	10^{18}	E	A	ά	alpha	O	o	omicron
Peta	10^{15}	P	B	β	beta	Π	ð	pi
Tera	10^{12}	T	´Γ	λ	gamma	p	ñ	rho
Giga	10^{9}	G		δ	delta	Σ	ó	sigma
Mega	10^{6}	M	E	ε	epsilon	T	ô	tau
Kilo	10^{3}	k.	Z	ξ	zeta	Y	õ	upsilon
deci	10^{-1}	d	H	η	eta	Φ	φ	phi
centi	10^{-2}	c	Θ	θ	theta	X	χ	chi
milli	10^{-3}	m	I	ι	iota	Ψ	ψ	psi
micro	10^{-6}	ì	K	κ	kappa	Ω	ω	omega
nano	10^{-9}	n	Λ	λ	lambda			
pico	10^{-12}	p	M	μ	mu			
femto	10^{-15}	f	N	ν	nu			
atto	10^{-18}	a	Ξ	ξ	xi			

II. UNITS OF MESAUREMENTS

Units of Length

1 metre	= 100 cm
1 cm	= 10 mm
1 mm	= 10^3 ìm
1 ìm	= 10^3 nm
1 nm	= $10^{2\,0}$ A* or 10^{-9} meter
	= $10^{7\,0}$ A
1 km	= 0.62 miles
1 mile	= 1.61 kilometers
5280 feet	= 1 mile
1760 yards (8 furlongs)	= 1 mile

Units of Area

1 sq. meter	= 10.76 sq. ft.
1 cent	= 436 sq. ft.
1 acre	= 0.405 hectare
1 hectare	= 2.47 acres
1 hectare	= 0.01 sq. km or
	= 10,000 sq meters.
4840 sq. yards	= 1 sq. mile
or 640 acres	= 1 sq. mile
9 sq. feet	= 1 sq. yard

1 light year $= 9.461 \times 10^{12}$ kilometers

                    1 cm = 0.3937 inch

                  1 inch = 2.5396 cm

                1 meter = 3.281 feet

                 1 feet = 0.3048 metre

Units of Volume

10^{-3} moles	= 1 millimole	
10^{-6} moles	= 1 micromole	
10^{-9} moles	= 1 nanomole	
10^{-12} moles	= 1 picomole	
1 litre	= 1000 ml	
1 ml	= 1000 µl	
1 ìl	= 1000 nl	
1 litre	= 0.264 gallon (USA)	
1 gallon	= 3.785 litres	
3.5 gallons	= 1 barrel	

Units of Mass

1 kg = 1000 gm		1 kg= 2.205 lbs. (USA)	
1 g = 0.001 kg		1 lb = 0.4536 kg	
1 mg = 1000 ìg		1 carat = 0.2 g (200 mg)	
1 g = 10^3 mg		1 ounce = 28.35 g	
1 g = 10^6 µg		1 g = 0.0353 ounce	
1 g = 10^9 ng		1 pound= 16 ounces	
1 g = 10^{12} pg		100 kg = 1 quintal	
1000 kg = 1 tonne			

Units of Depth

6 feet	= 1 fathom
120 fathoms	= 1 cable length
7.5 cable lengths	= 1 mile

Units of Speed

1 knot	= 1 nautical mile/hr
6080.2 feet	= 1 nautical mile
1 mach	= 1080 km/hr

Units of Temperature

$$^{o}C = (^{o}F - 32)\ 5/9$$

$$^{o}F = (^{o}C \times 9 / 5) + 32$$

$$^{o}K = (^{o}C + 273)$$

Are you overweight?

Body Mass Index is used to find out whether a person is overweight. In other words, BMI is the height to weight ratio used to measure adult heft and calculated as follows:

BMI = Weight in kgs / height in meters/ height in meters again (Up to 25 = Normal; between 25 and 30 = Overweight; 30 and above = obese)

Second indication of overweight is *Waist* to *Hip ratio*. That is, Waist measurement in cms / Hip measurement in cms (Should be < 0.9 for men and < 0.8 for women).

Some Physical Constants

Name	Symbol/ Abbreviation	Value
Faraday constant	F	96,485 c/mol.
Gas constant	R	1.987 cal/mol/K [1 cal = 4.184 J].
Planck's constant	*h*	1.584×10^{34} cal.K

Boltzmann constant	k	3.298×10^{-24} cal/K
Avagadro's number	N	6.023×10^{23} molecules
Curie	Ci	3.7×10^{10} dps or (2.22×10^{12} dpm)
Bequrrel	Bq	1 dps
Atomic Mass Unit (Dalton*)	amu	1.661×10^{-24} g
Velocity of light (in vaccum)	c	3×10^{8} m/sec.
Speed of sound	–	330 m/sec

*Dalton – a unit of mass very nearly equal to that of hydrogen atom or precisely equal to 1.0000 on Atomic Mass Unit **kiloDalton (kDa)** – unit of mass equal to 1000 daltons

calorie – one calorie (cal) is equivalent to the amount of heat energy required to raise the temperature of one gram of water from 14.5° C to 15.5° C. One **Kilocalorie** is equal to 1000 cal. One kilocalorie is enough energy to boil 12 ml of water. One **kilojoule*** is the energy to boil 2.9 ml of water or enable a person weighing 70 kg to climb eight steps.

(*an alternative unit of energy is the **Joule (J)** which is equal to 0.239 cal)

STANDARD pH VALUES

				Primary Standards				Secondary Standards	
Temperature	Bicarbonate-Carbonate Tartrate	Citrate	Phthalate	Calcium Phosphate	Phosphate	Borax	Bicarbonate-Carbonate	Tetroxalate	Hydroxide
°C	(Saturated)	(0.05M)	(0.05M)	(1:1)	(1:3.5)	(0.01M)	(0.025M)	(0.05M)	(Saturated)
0			4.003	6.982	7.534	9.460	10.321	1.666	
5			3.998	6.949	7.501	9.392	10.248	1.668	
10			3.996	6.921	7.472	9.331	10.181	1.670	
15			3.996	6.898	7.449	9.276	10.120	1.672	
20			3.999	6.878	7.430	9.227	10.064	1.675	
25	3.557	3.776	4.004	6.863	7.415	9.183	10.014	1.679	12.454
30	3.552		4.011	6.851	7.403	9.143	9.968	1.683	
35	3.549		4.020	6.842	7.394	9.107	9.928	1.688	
37			4.024	6.839	7.392	9.093			
40	3.547		4.030	6.836	7.388	9.074	9.891	1.694	
45	3.547		4.042	6.832	7.385	9.044	9.859	1.700	
50	3.549		4.055	6.831	7.384	9.017	9.831	1.707	
55	3.554		4.070					1.715	
60	3.560		4.085					1.723	
70	3.580		4.12					1.743	
80	3.609		4.16					1.766	
90	3.650		4.19					1.792	
85	3.674		4.21					1.806	

Dissociation Constants of Acids and Bases

Acids	Ka	Type	Bases	Kb	Type
HCN	4×10^{-10}		H_2NCONH_2	1.5×10^{-10}	
H_3BO_3	5.8×10^{-10}		$C_6H_5NH_2$	4.6×10^{-10}	v. weak
C_6H_5OH	1.3×10^{-10}	v. weak	NH_4OH	1.8×10^{-5}	
H_2S	1.1×10^{-7}		$(CH_3)_3N$	7.4×10^{-5}	
CH_3COOH	1.78×10^{-5}		CH_3NH_2	4.4×10^{-4}	weak
HCOOH	1.8×10^{-4}		$C_2H_5NH_2$	4.3×10^{-4}	
HF	6.7×10^{-4}	weak	$Ca(OH)_2$	3.7×10^{-3}	moderately strong
HNO_2	4.5×10^{-4}		NaOH	>10	
HCl	>10		KOH	>10	strong
HNO_3	>10	strong			
H_2SO_4	>10				
$HClO_4$	>10				

THE MEANING OF K

A constant of a reaction (K) defines the ratio of the concentrations of the products (e.g. dissociated ions) and the reactants (e.g. un-dissociated molecules).

For example, the dissociation constant K of a molecule HA is given as

K = [H+][A-]/[HA]. Square brackets denote concentration. Apparently this originated from Arrhenius's theory of electrolytic dissociation (among others): 'the molecule of an electrolyte can give rise to two or more electrically charged atoms or ions], and from the Law of Mass Action (by two Norwegians, Gulberg & Waage). If the dissociation happens in a solvent, at saturation, we can call the constant, solubility constant. The dissociation process might be endothermic or exothermic.

Let us look at a common dissociation: the autoprotolysis of pure water

The following equation describes the reaction of water with itself (called autoprotolysis):

$H_2O + H_2O <==> [H_3O^+ + OH^-$. The equilibrium constant for this reaction is written as follows: K = $[H_3O^+]$ OH-] / [H_2O][H_2O]. However, in pure liquid water, $[H_2O]$ is a constant, and together with K produces Kc.

Thus Kc = $[H_3O^+]$ $[OH^-]$. This constant Kc, also known as Kw, is called the water autoprotolysis constant or water autoionization constant. It is assumed to be obtained in dilute solutions, and may not strictly apply to concentrated solutions. Many factors such as temperature and the nature of solute would affect this equilibrium. For example, while Kwhas been shown to be 1.011 x 10-14 at 25 °C (generally, a value of 1.00 x 10-14 is used), Kw = 10-12.3 at 100 °C.

The equation above shows that concentrations of H_3O^+ and OH^- are in the molar ratio of one-to-one, and so $[H_3O^+] = [OH^-]$. Therefore, $[H_3O^+]$ and $[OH^-] = vKw$, which is 10^-7 M in pure water at 250C. This leads to important models of acids and bases, not least the pH scale.

Properties of commonly used commercial acids and bases(AR grades)

ACID/BASE	MW	SPECIFIC GRAVITY	ASSAY PERCENT (W/W)	GRAM/LITRE	NORMALITY	MOLARITY (M)
HCl	36.46	1.18	35.5	418.9	11.49	11.49
H_2SO_4	98.08	1.84	98.0	1,803.2	36.77	18.39
HNO_3	63.00	1.42	70.0	910.0	15.80	15.80
H_3PO_4	98.00	1.75	88.0	1,540.0	47.14	15.71
$HClO_4$	100.48	1.54	61.0	939.4	9.35	9.35
CH_3COOH	60.05	1.049	99.7	1,045.9	17.42	17.41
NH_3	17.03	0.910	25.0	227.5	13.36	13.36
NH_4OH	35.0	0.90	28.0	251.0	14.80	14.80

Dielectric constants of some common acids

Solvent	Dielectric constant at 20° C
Hexane	1.9
Benzene	2.3
Diethyl ether	4.3
Chloroform	5.1
Acetone	21.4
Ethanol	24.0
Methanol	33.0
Water	78.0 (80 at 0° C, 55 at 100° C
Hydrogen cyanide	116.0

Note: Higher Dielectric constant indicates higher polarity of solvent

Conversion of transmittancy (T%) to absorbancy (ABS = 2logT)

T(%)	Abs	T(%)	Abs	T(%)	Abs	T(%)	Abs
100	0.000	75	0.125	50	0.301	25	0.602
99	0.004	74	0.131	49	0.310	24	0.620
98	0.009	73	0.137	48	0.319	23	0.638
97	0.013	72	0.143	47	0.328	22	0.658
96	0.018	71	0.149	46	0.337	21	0.678
95	0.022	70	0.155	45	0.347	20	0.699
94	0.027	69	0.161	44	0.357	19	0.721
93	0.032	68	0.168	43	0.367	18	0.745

(Contd.)

92	0.036	67	0.174	42	0.377	17	0.770
91	0.041	66	0.181	41	0.387	16	0.796
90	0.046	65	0.187	40	0.398	15	0.824
89	0.051	64	0.194	39	0.409	14	0.854
88	0.056	63	0.201	38	0.420	13	0.886
87	0.061	62	0.208	37	0.432	12	0.921
86	0.066	61	0.215	36	0.444	11	0.950
85	0.071	60	0.222	35	0.456	10	1.000
84	0.076	59	0.229	34	0.469	09	1.046
83	0.081	58	0.237	33	0.482	08	1.097
82	0.086	57	0.244	32	0.495	07	1.155
81	0.092	56	0.252	31	0.509	06	1.222
80	0.097	55	0.260	30	0.532	05	1.301
79	0.102	54	0.268	29	0.538	04	1.398
78	0.100	53	0.276	28	0.552	03	1.523
77	0.114	52	0.284	27	0.569	02	1.699
76	0.119	51	0.292	26	0.585	01	2.000

Note: Readings of Absorbency values greater than 1 are generally ignored in measurements unless, special precautions are taken, because when Absorbency = 1, only 10% of the light is transmitted. For Absorbencies above>0.5, the readings are too close to get accurate readings. Therefore, read the Transmittancy and find out the corresponding Absorbency from the above Table.

Table - The Electromagnetic Spectrum

Electromagnetic Radiation	Wavelength (nm)	Energy of photons	Effect on molecules
Television waves $(10^3 - 10^5 \text{ cm})$	$10^9 - 10^{11}$	Low	-
Radio waves $(1 - 10^6)$	$10^6 - 10^9$		-
Radar	$10^6 - 10^{12}$		Spin orientations under Magnetic field eg.,NMR, ESR
Microwave $(0.1 \text{ cm} - 1 \text{ cm})$	$10^5 - 10^6$		Vibrations and / or rotations bond stretching or bending
Infrared region Far / Near	$2 \times 10^3 - 4 \times 10^5$ / $780 - 2 \times 10^3$		Vibrations and / or rotations, bond stretching or bending
Visible range Red / Orange / yellow / Green / Blue / Violet	$620 - 780$ / $590 - 620$ / $545 - 590$ / $490 - 545$ / $430 - 490$ / $490 - 430$		Valence shell electronic transition

(*Contd.*)

UV-rays	Long 300 – 390	Inner shell electronic
	Short 200 – 300	transition
	V. short 20 – 200	
X- rays	Soft 0.1 – 20	Nuclear transition
	Hard 0.0005 – 0.1	
Gamma rays	0.0001 – 0.14	High

pK values of compounds, commonly used in biological buffers

ACID/BASE	pK VALUES at 25° C
Acetic Acid	4.74
Ammonia (Ammonium)$^+$	9.26
Barbitone	7.98
Barbituric acid	4.98
Bicine*	8.35
Boric acid	9.23
Carbonic acid	6.37 and 10.25
Citric acid	3.11, 4.76 and 6.41
Formic acid	3.75
Glycine	2.35 and 9.68
Glycylglycine	8.40
HEPES*	7.60
Imidazole	6.95
MES*	6.15
MOPS*	7.20
PIPES*	6.80
Phosphoric acid	2.10, 7.20 and 12.30
Phthalic acid	2.90 and 5.51
Pyridinium	8.64
Succinic acid	4.19 and 5.57
Sulphuric acid	0.40 and 1.92
Tartaric acid	2.96 and 4.16
TES*	7.50
Tricine*	8.15
Tris*	8.14
Vernol (Sodium diethyl barbiturate)	8.00
Versene (EDTA)	2.0, 2.7, 6.2 and 10.3

Bicine*	: N,N-bis (2-Hydroxyethylglycine)
EDTA*	: Ethylenediamine tetraacetic acid
HEPES*	: N-(2-Hydroxyethyl piperazine-N^1-(2-ethanesulphonate)
MES*	: Morpholino Ethane Sulphonate
MOPS*	: Morpholino propane sulphonate
PIPES*	: Pipeerazine N-N^1-bis(2-ethanesulphonate)
TES*	: N-Tris (hydroxymethyl) 2-amino Ethane Sulphonate
Tricine*	: N-Tris (hydroxymethyl) methylglycine
Tris*	: N-Tris(hydroxymethyl) aminomethane

pK and pI values of amino acids in aqueous solution at 25 ° C

Amino acids	Symbol	MW	pK1	pK2	pK3	pK4	pI	Properties
1.Alanine	Ala,R	89.1	2.35	9.87	—	—	6.11	HP
2. b-Alanine	—	89.1	3.55	10.24	—	—	6.90	HP
3. Arginine	Arg, R	174.2	2.01	9.04	12.48	—	10.76	BA
4. Aspergine	Asn, N	132.1	2.14	8.72	—	—	5.43	AC
5. Aspartic Acid	Asp, D	133.1	2.10	3.86	9.82	—	2.98	AC
6. Cysteine	Cys, C	121.2	1.92	8.37	10.70	—	5.15	PO
7. Cystine	—	240.3	1.04	2.05	8.00	10.25	5.02	PO
8. Glutamic Acid	Glu, E	147.1	2.10	4.07	9.47	—	3.08	AC
9. Glutamine	Gln, Q	146.1	2.17	9.13	—	—	5.65	PO
10.Histidine	His, H	155.2	1.77	6.10	—	9.18	7.64	HP
11. Isoleucine	Ile, I	131.2	2.32	9.77	—	—	6.04	HP
12. Leucine	Leu, L	131.2	2.33	9.74	—	—	6.04	HP
13. Lysine	Lys, K	146.2	2.18	8.95	10.53	—	9.47	HP
14. Methionine	Met, M	149.2	2.13	9.28	—	—	5.71	HP
15. Phenylalanine	Phe, F	165.2	2.20	9.31	—	—	5.76	HP
16. Proline	Pro, P	115.1	2.00	10.60	—	—	6.30	PO
17. Serine	Ser, S	105.1	2.19	9.21	—	—	5.70	PO
18. Threonine	Thr, T	119.1	2.09	9.10	—	—	5.60	PO
19. Tryptophan	Trp, W	204.2	2.38	9.39	—	—	5.88	HP
20. Glycine	Gly, G	75.1	2.35	9.78	—	—	6.06	PO
21. Tyrosine	Tyr, Y	181.2	2.20	9.11	10.07	—	5.63	HP
22. Valine	Val, V	117.1	2.29	9.74	—	—	6.02	HP

(Adapted from Fluka Chemika Bio Chemica Analytica Catalogue. (1996)

Prop= Property; **AC**=Acidic; **BA**=Basic; **HP**=Hydrophobic; **PO**=Pair

How to find out the Logarithm and Antilogarithm of Numbers

a)The logarithm of any number contains *two parts*. One before the decimal is called *characteristic* and the decimal part is called *mantissa.* It is only for the mantissa we find the Logarithm from the Logarithmic table and the characteristic is found by the following rule i.e., it is always *less by unity* from the number of digits before the decimal point, if any, in a number. For example, in the number 980.00, there are 3 digits before the decimal point and hence the characteristic is 3 – 1 = 2. In the number 0.486, the number of digits before the decimal is 0 and hence the characteristic is 0 -1 = -1. The characteristic can be negative (in certain cases as in the second example) but the mantissa is always positive.

To find out the logarithm of a number, say, 3.14 (i.e. ð) which is most common factor in calculations, first find out its characteristic. Its characteristic is 1 -1 = 0. Mantissa of a number is same irrespective of the position of the decimal point and hence the mantissa of 314 will be same as that of 3.14. Therefore, the **Log of 3.14 = 0.4969 (Similarly, for 31.4, the logarithm is 1.4969)**

Antilogarithm is the *inverse of logarithm.* Proceed for decimal part exactly as in part (a) and for mantissa find out the antilogarithm using Antilogarithm table. For example, to find out the antilogarithm of 0.258, the number of digits before the decimal point is 0, then the *characteristic* is 0 + 1 = 1. (In antilogarithm, it will be always *more by unity* from the number of digits before the decimal point). Now go to Antilogarithm table and find out the number in *row 0.25* and *column 8,* which is 1811. Therefore, the **antilogarithm of 0.258 is = 1.811** (*as the characteristic is 1, the decimal point is placed after the first number*). **(Similarly, for 1.258, the antilogarithm will be 18.11)**

Table: Usage of Amino Acids:

Ala	GCU/C/A/G	Pro	CCU/C/A/G
Arg	CGU/C/A/G & AGA/G	Ser	UCU/C/A/G & AGU/C
Asp	GAU/C	Thr	ACU/C/A/G
Asn	AAU/C	Trp	UGG
Cys	UGU/C	Tyr	UAU/C
Glu	GAA/G	Val	GUU/C/A/G
Gln	CAA/G	Term	UAA/G/UGA
Gly	CGU/C/A/G		
His	CAU/C		
Ile	AUU/C/A		
Leu	CUU/C/A/G & UUA/G		
Lys	AAA/G		
Met	AUG		
Phe	UUU/C		

N.B: Single codon for Met & Trp, Three codons for Ile, Six codons for Arg, Leu & Ser

Table– Absorption maxima and molar absorption coefficients ($\acute{O}_M$) for Amino acids and Bases of Nucleic Acid at neutral pH.

Amino acids/ bases	$\ddot{e}_{max}$ (nm)	$\acute{O}$ at $\ddot{e}_{max}$ ($M^{-1} cm^{-1}$) x 10^{-3}
	280.0	5.6
Trp (204)	219.0	47.0
	274.0	1.4
	222.0	8.0
Tyr (181)	193.0	48.0
	257.0	0.2
Phe (165)	206.0	9.3
	188.0	60.0
His (155)	211.0	5.9
Cys (121)	250.0	5.9
Adenine (135.1)	260.5	13.4
Guanine (151.5)	275.0	8.1
Cytosine (111.1)	267.0	6.1
Uracil (112.1)	259.5	8.2
Thymine (126.1)	264.5	7.9
DNA	258.0	6.6
RNA	258.0	7.4

*Other amino acids do not absorb light significantly (Molecular weight is given in brackets)

FORMULA WEIGHTS FOR COMMONLY USED CHEMICALS

Chemical Formula Weight (grams/Mole)

- Adenosine triphosphate (ATP)   605.2
- Ammonium Acetate   77.08
- ß-mercaptoethanol   78.13
- Boric Acid 61.83
- Dithiothreitol (DTT)   154.3
- EDTA-disodium salt (Na2EDTA)   372.2
- Ethidium Bromide (EtBr)   394.3
- Glucose (Dextrose)   180.2
- Hydrochloric Acid (conc. HCl = 12.1 N)   36.5
- Potassium Chloride (KCl)   74.56
- Potassium Acetate (KOAc)   98.15
- Magnesium Chloride (MgCl2 * 6 H2O)   203.31
- Magnesium Acetate (Mg(OAc)2 * 4 H2O)   214.46
- Magnesium Sulfate (MgSO4 * 7 H2O)   246.5
- Spermidine Tetrahydrochloride   254.6
- Sodium Acetate (NaOAc)   136.08
- Sodium Chloride (NaCl)   58.44
- Sodium Citrate (citric acid, sodium salt)   294.1
- Sodium Hydroxide (NaOH)   40.0
- Sodium Phosphate, monobasic (NaH2PO4)   120.0
- Sodium Phosphate, dibasic (Na2HPO4)   142.0
- Trizma Base   121.1

MISCELLANEOUS FACTS AND UNITS

- Avogadro's Number: $6.02 \times 10e23$
- 1 bp DNA = 660 daltons
- 1 Kb duplex DNA = $6.6 \times 10e5$ daltons
- 1 Kb duplex DNA = 333 amino acid coding capacity = 37,000 dalton protein
- 1 uM = 1 pmole/ul
- pM = 10e– 12 M
- 1 ug/ml of a 1 Kb nucleic acid fragment = 3 nM fragment ends
- a 1 x 10e6 dalton restriction fragment = 1 ug/pmole of fragment
- bacteriophage lambda = $30 \times 10e6$ daltons, = 30 ug/ pmole lambda
- 1 bacteriophage lambda plaque contains ~ 10e6 phage particles
- a lambda lysate has approximately $2 \times 10e10$ pfu /ml
- 1 O.D. of duplex DNA; A260 of 1.0 = 50 ug/ml
- 1 O.D. single-stranded RNA; A260 of 1.0 = 40 ug/ml
- 1 O.D. single-stranded DNA; A260 of 1.0 = 33 ug/ml
- O.D.660 *E. coli* cells $\times 5$ = cells/ml $\times 10e8$
- haploid human genome = $3 \times 10e9$ bp
- haploid human genome = 33 Morgans in length (3300 cM)

- haploid yeast (*S. cerevisiae*) genome = $1.5 \times 10e7$ bp
- *E. coli* genome = $4.72 \times 10e6$ bp
- wild type lambda genome = $4.85 \times 10e4$ bp
- Density of duplex DNA = ([molar fraction G + C] + 14.19) / 8.63 ~ 1.7 g/ml
- Density of RNA ~ 1.9 g/ml
- Density of lambda phage ~ 1.5 g/ml
- Density of protein ~ 1.3 g/ml
- Density of carbohydrate ~ 1.5 - 2.0 g/ml
- 1 uCi = $7 \times 10e6$ cpm (Cherenkov radiation) = $2.2 \times 10[6]$ dpm
- e32P half life = 14.3 days
- e35S half life = 87.4 days

ABBREVIATIONS AND ACRONYMS

- A260: Absorbance at 260 nm wavelength (UVlight)
- ABI: Applied Biosystems, Inc.
- ALARA: as low as reasonable achievable
- amp or AMP: ampicillin
- APS: ammonium persulfate
- ATCC: American Type Culture Collection
- ATP: adenosine triphosphate
- BCS: bovine calf serum
- BIOS: The BIOS Corporation
- BMB: Boehringer Mannheim Biochemicals
- ßme, 2-ME: Beta-mercaptoethanol
- bp: base pairs
- BRL: Bethesda Research Laboratories
- BSA: bovine serum albumin
- (CA)n: dC*dA dinucleotide repeat
- CGM: Center for Genetics in Medicine (WUMS)
- CEPH: Centre d'Etude Polymorphisme Humain
- CHEF: Contour-clamped homogeneous electric fields
- CIP: calf intestinal alkaline phosphatase
- cm: centimeters
- cM: centimorgans
- cntl, CTRL: control key
- CO2: carbon dioxide
- cpm: counts per minute
- CRI: Collaborative Research, Inc.
- CSA: cyclosporin A
- DAPI: 4,6-diamidino-2-phenylindole-dihydrochloride
- ddH_2O: distilled & deionized water
- ddNTP: dideoxyribonucleoside triphosphate
- DGGE: denaturing gradient gel electrophoresis
- dH2O: deionized or distilled water
- DMEM: Dulbecco's modified eagle medium (tissue culture)
- DMSO: dimethyl sulfoxide
- dNTP: deoxyribonucleoside triphosphate (usually one of dATP, dTTP, dCTP, dGTP)
- dpm: disintegrations per minute
- DTT: dithiothreitol
- EBV: Epstein Barr virus
- EDTA: ethylenediaminetetraacetic acid (powder is a disodium salt)
- EIU: electronic interface unit (Biomek)
- EMBL: European molecular biology laboratory
- EtBr: ethidium bromide
- EtOH: ethanol
- FBS: fetal bovine serum
- FF: father's father
- FITC: fluorescein isothiocyanate
- FM: father's mother
- FW: formula weight
- GTL: G-banded chromosomes using trypsin and Leishman's stain
- HEPES: N-2-hydroxyethylpiperazine-N'-2-ethanesulfonic acid
- HET: heterozygosity
- HGM: human gene mapping
- IM: informative meioses

- IPTG: isopropylthio-ß-D-galactoside
- ISCN: International System for Human Cytogenetic Nomenclature
- kan: kanamycin
- Kb, kb: kilobases
- Kg, kg: kilograms
- LCP: Linkage Control Package
- mA: milliamperes
- uCi: microCuries
- MEM: minimal essential media (tissue culture)
- MeOH: methanol
- MF: mother's father
- MIM#: Mendelian Inheritance in Man Number
- ug: micrograms
- ul: microliters
- mg: milligrams
- ml: milliliters
- MM: mother's mother
- MW: molecular weight
- ng: nanograms
- nm: nanometers
- OD260: Optical density at 260 nm wavelength (UVlight); = A260
- oligos: Oligonucleotides
- OS: operating system
- e32P: radioactive phosphorus isotope
- PAGE: polyacrylamide gel electrophoresis
- PBS: phosphate buffered saline
- PCR: polymerase chain reaction
- PEG: polyethylene glycol
- PFGE: pulsed field gel electrophoresis
- pfu: plaque forming units
- PHA: phytohemagglutinin
- PI: propidium iodide
- PIC: polymorphism information content index
- PMSF: phenylmethylsulfonyl fluoride
- RAM: radioactive materials
- RFLP: restriction fragment length polymorphism
- rpm: revolutions per minute
- RPMI-1640: Roswell Park Memorial Institute medium (tissue culture)
- RSO: radiation safety officer
- R.T.: room temperature
- e35S: radioactive sulfur isotope
- SDS: sodium dodecyl sulfate
- Solution A (tissue culture support center): Dulbecco's Ca++/Mg++ free medium
- STS: sequence tagged site
- TA, TAE: Tris-acetate buffer
- TB, TBE: Tris- borate buffer
- TC: tissue-culture
- TE: Tris- EDTA buffer
- TEMED: N,N,N',N'-tetramethylethelenediamine
- Tet: tetracycline
- TSC: Tissue Support Center (WUMS)
- U: units
- UV: ultra-violet light
- V: volts
- VNTR: variable number of tandem repeats
- v/v: volume per volume
- W: watts
- WBC: white blood cells
- WUMS: Washington University Medical School
- w/v: weight per volume
- X-gal: 5-bromo-4-chloro-3-indolyl-beta-D-galactoside
- YAC: yeast artificial chromosome

COMPOSITION AND PREPARATION OF MEDIA

Selective Methods

There are three different methods for subdividing a complex mixture and mixed population of bacteria (microorganisms) into groups of similar ones, called selective methods, by which one can achieve an increase in the relative number of a particular species in respect to others by discouraging or killing undesirable species. These methods are (1) Chemical, (2) Physical, and (3) Biological, as summarized in the following table after Pelczar *et al. (1986).*

CHEMICAL METHODS

1. Enrichment media: Use of special nutrient sources
In this method specially formulated culture media are used, i.e. A single source of some elements is used that can only be utilized by the kind of bacteria of interest. For eg. *Salmonella typhi* can grow in a selenium supplement medium in place of phosphorus. While other bacteria growth are inhibited in the presence of Selenium.

2. Use of dilute media
Most bacteria grow in media containing minimum threshold level of nutrients, yet some bacteria can survive and grow even though these nutrients are present in a very low concentration (e.g. *Coulobactor*). Thus in a nutrient deficient culture all but the one required bacteria will die leaving a pure culture of the desired one.

3. Use of selective media containing inhibitory or toxic chemicals
Media is supplemented with low concentration of some specific chemicals like dyes, bile salts or antibiotics. Thus by using a medium containing crystal violet (the dye used for Gram staining; inhibits Gram –positive bacteria) and sodium deoxycolate (a bile salt that inhibit growth of non-intestinal bacteria), Gram–negative intestinal bacteria (e.g. *Shigella sp.*) can be selected.

PHYSICAL METHODS

1. Application of Heat
For selecting endospore-forming bacteria, a preheated (up to 80°C for 10min.) mixculture can be used as an inoculum for a fresh culture. Heat destroys all but the endospores, which eventually give rise to a pure culture.

2. Using different temperature range
Incubation temperature (between 0°C to 5°C preferentially select psycrophilic / psychrotrophic bacteria in preference to the thermophilic bacteria. Identically higher temperature range (50°C to 55°C) can select thermophiles in preference to psycrotrophics.

3. Alteration of pH of the medium
For selecting acid sensitive bacteria, such as lactobacilli, in preference to acid sensitive forms, a low pH-medium can be employed. Similarly alkali-tolerant bacteria can be preferentially selected on a high pH medium.

BIOLOGICAL METHOD

Human pathogenic bacteria can be isolated by infecting some susceptible laboratory animals with sample of mixculture containing the target bacterium as indicated by the disease symptoms. The target pathogen will thrive in the susceptible animal while others will be inhibited or killed by the immune response of the animal. The target bacterium can be isolated from the serum of the animal used.

Bacterial Media

LB Medium

Yeast extract	-	5 g
Tryptone	-	10 g
NaCl	-	10 g
DH_2O	to 1000 ml (adjust the pH to 7.2 with 5N KOH)	

M9 Medium (10X)
  Solution 1:

Na_2HPO_4	-	6.0 g	
KH_2PO_4	-	3.0 g	
NaCl	-	0.5 g	
NH_4Cl	-	1.0 g	
dH_2O to	-	100.0 ml	

Solution II: 1 M $MgSO_4.7H_2O$. (10 ml)
Solution III: 0.1 M $CaCl_2$. (10 ml)
Solution IV: 40% glucose in dH_2O. (100 ml)

Distilled H_2O. (autoclave required volume)
All solutions are autoclaved separately. *The final medium is prepared as follows:*
To 90 ml sterile dH_2O, add 10 ml M9, 0.1 ml $MgSO_4$, 0.1 ml $CaCl_2$ and 5 ml glucose solutions.
Yeast media:

YEPD

Yeast Extract	-	10 g
Peptone (M)	-	15 g
dH_2O	-	950 ml

Prepare a 40% solution of Glucose stock (50 ml), autoclave separately and
Add to the above medium to 2% final concentration.
(*For inoculum preparations YEPD is used*).

M9 Modified Medium (10X)

*For labeling experiments, the following modified M9 synthetic medium is
Used.*

Modified M9 (10X) is prepared by adding 4 g ammonium sulphate, 1 g yeast
nitrogen base (without amino acids and ammonium sulphate) and 0.1 ml Vogel's trace element stock
solution to 10X M9. pH is adjusted to ~ 6.0 and autoclaved. This solution is diluted to 1:10 in sterile
water and autoclaved glucose stock solution is added to 2% just before inoculation.
Fungal Medium (Vogel's medium):

Stock solution –1 (Vogel's (M) Salt solution, 50 N)

1. Distilled H_2O -   775 ml
2. Trisodium Citrate   -   125 g
3. KH_2PO_4 -   250 g
4. NH_4NO_3 -   100 g
5. $MgSO_4.7H_2O$ -   10 g
6. NaCl   -   25 g

Make up final volume to 1000 ml. Add a few drops of chloroform, shake well and store at room
temperature. The pH is ~ 6.0. The solution can also be autoclaved and stored.

Stock solution- 2 (Vogel's (M) Trace elements stock)
1. dH_2O -   95 ml
2. Citric acid   -   500 mg

3. $ZnSO_4.7H_2O$ -   500 mg
4. $Fe(NH_4)_2SO_4.6H_2O$ -   500 mg
5. $CuSO_4.5H_2O$ -   25 mg
6. $CoCl_2$ -   25 mg
7. $MnSO_4.H_2O$ -   50 mg
8. H_3BO_3 -   50 mg
9. $(NH_4)_6Mo_7O_{24}.4H_2O$ -   50 mg

Mix 1.5 ml of 1M $CaCl_2$ slowly in drops. Make up the final volume to 100 ml. Add a few drops of chloroform, shake well and store at 4° C

Stock solution 3 : ($CaCl_2$ solution)

Five g $CaCl_2$ in 10 ml dH_2O (50%)

Stock solution 4 (Vitamins)

1. d-Biotin   -   0.5 mg
2. Myo Inositol   -   200.0 mg
3. Calcium Pantothenate   -   20.0 mg
4. Pyridoxine HCl   -   20.0 mg
5. Thiamine   -   20.0 mg
6. dH_2O to   100 ml

PREPARATION OF FINAL MEDIUM

Solution 1 (50 N Salt solution)   -   20.0 ml
Solution 2 (Trace element solution)   -   1.0 ml
dH_2O to 1000 ml

The pH is ~ 6.0; if not adjust with 5 N NaOH or 1 N HCl. Add carbon source to 2% - 3% to the above medium and sterilize by autoclaving. This modified Vogel's medium supports growth of most of fungal strains. (For mycelial incoculum preparation. 0.2% mycological peptone is also added before autoclaving the final medium.

Note

* $CaCl_2$ stock solution is prepared separately, autoclaved and added to the sterile medium just before inoculation at 0.4 ml/litre.

**Most fungi do not require vitamins at all. If required vitamin stock solution is prepared, filter sterilized and added at 1ml/litre. *Neurosporas* require only d-Biotin.

Quick Reference to Abbreviations and Symbols

At.wt.	Atomic weight	N	Normal (concentration)
C	Concentration	N	Avogardo's number
Conc, C	Concentration	N	Number of equivqlents per mole
Equiv.	Equivalent	V, vol	Mole
Equiv wt.	Eqivqlent weight	Wt	Volume
Fw	Formula weight	w/w	Wight
g	Gram	w/w	Weight-to-weight ratio
L	Liter		Weight-to-volume ratio
M	Molarity	μ	Density
mol	Mol	[ ]	Micro
Mol wt	Molecular weight		Molar concentration of substance in bracket

Quick Reference to Unit Conversion

	Gram (g)	
Unit	Symbol	Equivalent (g)
picogram	pg	$1\ pg = 10^{-12}$
nanogram	ng	$1\ ng = 10^{-9}$
microgram	μg	$1\ \mu g = 10^{-6}$
milligram	mg	$1\ mg = 10^{-3}$
kilogram	kg	$1\ kg = 10^{3}$

	Mole (mol)	
Unit	Symbol	Equivalent (g)
picomole	pmol	$1\ pmole = 10^{-12}$
nanomole	nmol	$1\ nmole = 10^{-9}$
micromole	mol	$1\ \mu mole = 10^{-6}$
millimole	mmol	$1\ mmole = 10^{-3}$

	Liter (l or L)	
Unit	Symbol	Equivalent (g)
nanoliter	nL	$1\ nL = 10^{-9}$
microliter	μL	$1\ \mu L = 10^{-6}$
milliliter	mL	$1\ mL = 10^{-3}$

CONVERSIONS

- or conversion of base pairs into molecular weight (in Daltons), multiply by 660, as average molecular weight of one base pair = 660 Daltons.
- For conversion of nucleotides into molecular weight (in Daltons), multiply by 330 as average molecular weight one nucleotides = 330 Daltons.

- For conversion of amino acids into molecular weight (in Daltons), multiply by 110 as average molecular weight for an amino acid = 110 Daltons. Therefore, a protein with 300 amino acids will be having a molecular weight of ~33,000 Daltons or 33kDa.

Quantification of DNA, RNA and Oligonucleotides

1 A_{260} Unit of double-stranded DNA	= 50 mg/ml
1 A_{260} Unit of single-stranded DNA	= 37 mg/ml
1 A_{260} Unit of Single-stranded DNA	= 40 mg/ml
1 A_{260} Unit of oligonucleotides	= 20 – 33 mg/ml

For pure DNA, A_{260}/A_{280} = 1.8 and for pure RNA A_{260}/A_{280} = 2.0

Protein markers for SDS-PAGE (in Daltons)

Thyroglobulin	- 300,000
Myosin (Rabbit muscle)	- 205,000
b-Galactosidase (*E.coli*)	- 116,000
Phosphorylase-a (Rabbit muscle)	- 97,000
Phosphorylase-b (Rabbit muscle)	- 97,400
Bovine serum albumin (BSA)	- 66,000
Ovalbumin (Chicken egg)	- 43,000
Carbonic anhydrase (bovine or human erythrocytes)	- 29,000
b-Lactalbumin (bovine milk)	- 18,400
Myoglobin (horse heart)	- 17,000
Lysozyme (Chicken egg white)	- 14,300
Ctochrome- C	- 11,700
Aprotinin (bovine lung)	- 6,500

Isoelectric pHs (pI) of enzymes

Pepsin	-1.0
Amyloglucosidase (*A.niger*)	-3.6
Glucose oxidase (*A.niger*)	-4.2
Urease (Jack bean)	-4.0
DNA polymerase (human lymphocytes)	-4.7
RNA polymerase II (human He La cells)	-4.8
Ovalbumin (chicken egg)	-4.8
b-Lactoglobulin-A (bovine milk)	-5.1
Tubulin (pig brain)	-5.5
Carboxypeptidase B (human pancreas)	-6.0
Carbonic anhydrase (bovine erythrocytes)	-6.1
Pyruvate kinase (rat muscle)	-7.5
Ribonuclease (bovine pancreas)	-9.3
Proteinase-K (*Tritirachium album*)	-8.8
Cytochrome C (horse heart)	-10.2
Lysozyme (hen egg white)	-11.1

Table: Properties of Sephacryl range of gel chromatography media
(by courtesy of Pharmacia Ltd.)

Sephacryl type	Wet bead diameter (mm)	Fractionation range	
		Proteins	Polysaccharides
Sephacryl S200	40–105	5×10^3–2.5×10^5	1×10^3 –8×10^4
Sephacryl S300	40–105	1×10^4–1.5×10^6	2×10^3 –4×10^5
Sephacryl S400	40–105	3×10^4–8×10^6	1×10^4 –2×10^6
Sephacryl S500	40–105	-	4×10^4 –20×10^6
Sephacryl S1000	40–105	-	$5 ´ 10^5$ –$<10^8$

Table - The Electromagnetic Spectrum

Electromagnetic Radiation	Wavelength (nm)	Energy of photons	Effect on molecules
Television waves ($10^3 – 10^5$ cm)	$10^9 – 10^{11}$	Low	-
Radio waves ($1 – 10^6$)	$10^6 – 10^9$		-
Radar	$10^6 – 10^{12}$		Spin orientations under Magnetic field eg.,NMR, ESR
Microwave (0.1 cm – 1 cm)	$10^5 – 10^6$		Vibrations and / or rotations bond stretching or bending
Infrared region — Far	$2 \times 10^3 – 4 \times 10^5$		Vibrations and / or rotations, bond stretching or bending
Infrared region — Near	$780 – 2 \times 10^3$		
Visible range — Red 620 – 780, Orange 590 – 620, yellow 545 – 590, Green 490 – 545, Blue 430 – 490, Violet 490 – 430, Long 300 - 390			Valence shell electronic transition
UV-rays — Short 200 – 300, V. short 20 – 200			Inner shell electronic transition
X- rays — Soft 0.1 – 20, Hard 0.0005 – 0.1			Nuclear transition
Gamma rays	0.0001 – 0.14	High	